1993

THE FRACTAL EXPLORER

Linda Garcia

D1546361

Dynamic
Press

Dynamic Press
Santa Cruz, CA

Edited by Celene de Miranda

Illustration and layout by Cory Levenberg

This book was created electronically on Apple® Macintosh Computers and NeXT® Computers.

Design, layout and artwork was produced using Aldus Pagemaker™, Microsoft Word™, FractaSketch™, MandelMovie™, Aldus Freehand™, Adobe Photoshop™, PixelPaint Professional™, CricketGraph™, Mathematica™, Fractal Contours™, LightspeedC™, and Turbo Pascal™.

Printed in the United States of America,   First Edition

ISBN 0-9628659-0-7   Dynamic Press   Santa Cruz   Berkeley

I dedicate this book to my brother David, who, if he remembers to breathe, can do anything.

The Fractal Explorer is published by Dynamic Press®, a group of mathematicians and developers closely associated with the fractal, chaos and dynamics research group at the University of California at Santa Cruz. The goal of Dynamic Press® is to further the understanding of mathematics and science in conjunction with advances in computational science. The group promotes creativity through shared contributions, a long standing Santa Cruz tradition.

The Designer Fractal™ series of publications and software was developed to enhance exploration of fractal geometry through visual interactive environment: a living text book. The Designer Fractal™ series includes the application "FractaSketch™" by Peter Van Roy at the University of California at Berkeley, "MandelMovie" by Michael Larsen at the University of Pennsylvania, and "Chaos" by Bernt Wahl at the University of California at Santa Cruz. The book *The Fractal Explorer* by Linda Garcia with contributions from Peter Van Roy, Michael Larsen, Celene de Miranda, Cory Levenberg and Bernt Wahl, describes the general aspects of Fractal Geometry, and is our latest addition to the Designer Fractal™ series. Linda's delightful and informal talks on fractals have brought appreciation and understanding to many explorers of this new field of mathematics, and her book is sure to do the same. Additions by others in our group are also planned, including an instructional video series and gallery posters.

Our group would also like to thank the many contributors who have helped in our projects: the mathematicians who shared much of their work with us and who were a great source of inspiration, the beta testers and proof readers whose input and "debuggery" made a final copy possible, and of course the countless people like yourself who helped and encouraged us. Without your help Dynamic Software & Press™ would have remained only an idea. Please direct comments, questions, and orders to Dynamic Software, P.O. Box 7534, Santa Cruz, CA, 95060.

Sincerely,

Bernt Wahl

Peter Van Roy

The Group from Dynamic Software

ACKNOWLEDGEMENTS

In addition to the enthusiasm and support from countless people, the following were especially instrumental to the research, writing, design and/or publication of The Fractal Explorer.

Special thanks to Cindy Nye, for the joint research, programming, preparation, discovery, and learning in 1987 and 1988 that went into the presentation that is the foundation of this book. Thanks to Dr. Michael Hortmann for his uninterrupted help with the same.

To Miles Sundermeyer, in addition to three years of foundation-setting, question-asking, and problem-solving, thanks for much work on the soon-to-be included section on Chaos. Thanks also to Jennifer Dunvan for work on the same. To Ron Eglash, thanks for helpful comments, suggestions, and (now) pending contributions to the application section of the book. To Sean Ballard thanks for help with experiments on Chaos at Silicon Systems Inc. in Santa Cruz.

Thanks to Jill Whitby for her help with resources. Thanks to Dave Garcia, great hunter and gatherer, for an enormous amount of footwork on research, help with the bibliography, and for the collection of various fractals, (assisted in 1988 by Rachel Herzig). Thanks to Susan Nishiyama for proofreading the text. Thanks to Angelique Sundermeyer for an excellent job on the artwork and design of the cover.

To The Group at Dynamic Software, much appreciation for the hard work in the design, format, and especially the original illustration of the text. Thanks to Dr. Peter Van Roy for the section "Growing Gardens from Fractal Seeds", for other numerous illustrations from FractaSketch™, and for his helpful comments and suggestions regarding the text. Thanks to Dr. Michael Larsen for the illustrations from MandelMovie™. Thanks to Cory Levenberg for the format and layout of the text, and for an enormous amount of work in designing and generating the illustrations. Thanks also to Valerie Lieu for her help with, and especially her patience, concerning the layout. Thanks to Dave Saber for "the sentence" and for allowing us to invade his apartment. Finally, thanks to Bernt Wahl for the appendices and for having asked in the first place. Without his request in October 1988, there would be no book.

Most of all, I would like to thank Celene de Miranda, friend and editor extraordinaire. Without a doubt, Celene is the only person in the world who really knows what it took to write this book. Without her expertise, unending encouragement, devotion and much personal sacrifice, this book would have remained trapped in some corner of my mind (or worse yet, for those of you who know me, in an undone pile of my laundry).

Many have been denied the real beauty of mathematics, but with her careful editing and attention to audience, the world of Fractal Geometry is now that much more accessible. It was an unforgettable experience, Celene. Thank you.

Last, I would like to thank my family and friends for their guidance, love, and support. For a quiet push to pursue what I loved, thanks to my parents Guadalupe and Virginia Garcia, and to their awesome children: John, Arlene, and David.

To Natalie, Bob and Dimitri, thanks for the years 1984, 1987, 1988, and their consequences, one of which is this book.

When I first began to learn about Fractal Geometry I was motivated by the sheer beauty of the images that I saw. Though I did not understand how these images were generated, or who was interested in them or why, they were certainly inviting, and I was intrigued. Then Peter Oppenheimer came to speak at the University of California, Santa Cruz, and a friend suggested that I go. I remember clearly that at first I understood little. But I was used to that; my major was mathematics, and I was convinced that it was my fate to sit in classrooms and take notes on lectures that no one (especially myself) understood. But this was different; Mr. Oppenheimer used pictures, and the word "accessible" came to mind. I sat in the back of the lecture hall, and in this mathematics presentation we watched, of all things, trees grow! I had never seen a tree grow. In fact I had never really spent any time even thinking about how a tree grows. Here, however, were some very life-like images that demonstrated the branching structure that underlies their growth.

I left the presentation and stopped dead in my tracks. It was just an oak tree, but I felt like I was looking at a tree for the first time in my life. The oak trees at the university were amazing at night, and lit from underneath by streetlamps, their structure seemed almost unreal. And now this structure was mathematical?! This made the tree even more beautiful, and I wondered why no one had ever told me about this before.

Then began stage one, question one: What is a fractal? This was in the spring of 1987, and from then until the spring of 1988 I spent every spare minute trying to answer that question. Along the way I found what always seemed to me small miracles of mathematics. I attended a special presentation by Benoit Mandelbrot, Heinz-Otto Peitgen and Richard Voss. I learned how to generate the Mandelbrot set, and I spent some time studying the work of Michael Barnsley of Georgia Tech.

Then began the great hunting and gathering quest, and I started collecting fractal articles, books, and pictures like a maniac. All in search of the answer to question one: what exactly is a fractal? All this was done with the help of visiting professor Michael Hortmann. After a year of searching, I finally asked him the question, "What is a fractal?" "A fractal," he said, "is undefined."

Oh.

Rather an anticlimactic answer to the question. And why a year to find out that I was searching in vain? The answer to that question became as interesting as the first, and with the continued help of Michael, I prepared to give a joint talk on fractals with a fellow student, Cindy Nye. Roughly one year after Peter Oppenheimer's talk, Cindy and I gave our own talk: Fractal Geometry Demystified.

Stage two, question two: What are fractals good for? This second question came quite naturally and the quest became to collect every article ever written, in every scientific journal, popular magazine, book or newspaper about the research being done concerning fractal applications. It may be an impossible task, but after two years, I still don't know that. However, I do know that even though I am literally surrounded by 200 articles, I'm about 600 articles shy and nowhere near being finished. In fact I am smack-dab in the middle of stage two, question two. I like this question even more than the first. I can't possibly do justice to all the research being done on fractals. My hope is to someday write an entire book that surveys this.

Stage three, question three: How are fractals related to the field of Chaos? In a nut shell, the fields of Chaos and Fractal Geometry are mathematical cousins. Chaotic systems settle down to shapes that have fractal dimension; these are called strange attractors, and strange attractors are fractal. The problem with nut shells of course, is that if you're really hungry you can open them. After a class on Chaos with Ralph Abraham, I now realize that my fractal universe is just a slice of this field of Chaos.

It's press time and I'm caught right in the middle of my quest to answer these questions. Three and a half years ago I had never heard of a fractal, and now there is this book. I guess it will continue to be the book that was "always never finished." But to say that I was finished would mean that I had finished learning. Which is not the case. It all seems very fitting, however, that considering the infinite worlds that lie within just one mathematical fractal, the quest of the fractal explorer is by definition never over.

"It's ironic that fractals, many of which were invented [by 19th century mathematicians] as examples of pathological behavior, turn out not to be pathological at all. In fact they are the rule in the universe. Shapes which are not fractal are the exception. I love euclidean geometry, but it is quite clear that it does not give a reasonable presentation of the world. Mountains are not cones, clouds are not spheres, trees are not cylinders, neither does lightning travel in a straight line. Almost everything around us is noneuclidean."

Benoit Mandelbrot

"We ought not to believe that the banks of the ocean are really deformed because they are not exact pyramids or cones; nor that the stars are unskillfully placed, because they are not all situated at uniform distance. These are not natural irregularities but with respect to our fancies only."

Richard Bentley 17th Century

Fractal Geometry is often introduced as *a geometry of nature.* Yet if we think about it, what else would geometry be about? Since the word itself comes from the Greek geometrain— to measure the earth- the term "Geometry of Nature" might seem redundant.

Yet it is not. Why? Because Euclid's definition, with its spheres, cones, right angles, and straight lines, holds a 2000 year monopoly on our meaning of the word geometry. From the time we are small children we are given the false impression that if a phenomenon does not fit within these strict definitions it is not geometric at all. Circles, triangles, rectangles, and lines were fine until we turned to look out the classroom window. We looked beyond the square window pane to the trees, mountains, and clouds outside and we believed that geometry ended there.

We believed this because mathematicians believed this. They looked out the window and then turned their backs as quickly as possible. Had nature with its rugged mountains, rough coastlines and non-spherical clouds failed to fit within the boundaries of Euclidean definitions? Or had scientists failed by disregarding forms and structures that were not smooth, predictable, or ordered? Given the bias of the human ego, it was of course nature that had failed, and ideas to the contrary were regarded as exceptions. As a result, students of mathematics were given the mistaken impression (and still are) that the Euclidean is the rule rather than the exception.

Bentley pointed out that as a result of the overzealous attempt to be linear, Newtonian, and Euclidean, our fancies have taken the word *irregular* and mistakenly assigned it to nature. Now the limitations of the Euclidean alphabet have surfaced. Fractal Geometry has stepped forward with mathematical tools capable of both measuring and simulating nature's complexity, and the term irregular must now be reassigned.

TABLE OF CONTENTS

INTRODUCTION

Fractals finally do justice in revealing to everyone, the layperson included, the beauty of mathematics.

FRACTAL GEOMETRY

Recently a subscriber to *Physics Today* complained that every third article submitted to the magazine was about fractals. What are these fractals? What is Fractal Geometry, and why is it being discussed in such a variety of circles? These questions are echoed by many. With even a brief introduction to these extraordinary objects, the fascination with the field is understandable.

Since its inception roughly 15 years ago one thing is clear— Fractal Geometry is exciting. As with many new subjects, those who are discovering this young branch of mathematics feel as if somebody has given them a new pair of glasses. The positive response is due to the fact that studying fractals is immediately gratifying. Not only are fractals beautiful, but in an uncanny sense, they seem quite familiar. They defy intuition with their infinite and repetitive worlds, yet they remain accessible. This surprises many because, as Benoit Mandelbrot points out, "in the past mathematicians have increasingly chosen to flee from nature by devising theories unrelated to anything we can see or feel" [Mand82]. Today Mandelbrot is credited with establishing the theory of fractals and for having expanded geometry to include and model nature exceptionally well. As shown in his book, *The Fractal Geometry of Nature*, and in other books such as Peitgen and Richter's, *The Beauty of Fractals*, fractals finally do justice in revealing to everyone the beauty of mathematics.

Through Fractal Lenses

This geometry of nature provides a new filter for viewing the world around us. With a basic introduction, the world of the reader is transformed. After people are introduced to fractals, they can use their new "glasses" to see the fractal qualities in a leaf, a tree, a mountain, ice on a window, or cracks in a sidewalk. As Michael Barnsley says in the preface of his book, *Fractals Everywhere*, there is a danger in reading further. Clouds, trees, ferns, and a host of other objects never quite look the same again. When exposed to the briefest explanation, people are surprised at their own excitement. Reactions range from the very common, "Look at all the fractals!" to "the straight line has become an absolute tyranny...drawn without thought or feeling; it is the line which does not exist in nature" [Peit86].

In addition to their sheer beauty, these fractals have practical applications that have surfaced in almost every field imaginable. With a delicate balance between "order" and "chaos," the infinite worlds that lie hidden in these descriptions of nature's complexity have served as an invitation to both hobbyists and scientists alike. Like mathematics in general, Fractal Geometry gives us a way to explain and enhance the world around us. Like a near-sighted child surprised by the magic of eyeglasses, once we begin to look at our world through the lenses of Fractal Geometry, we wonder how we could have possibly done without it.

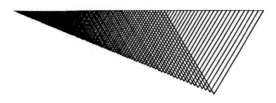

HISTORICAL BACKGROUND

CLASSICAL FRACTALS

The scientific community has not always embraced the ideas behind Fractal Geometry. In fact, just the opposite is true. Its foundations were established on the outskirts of traditional mathematics and its history contains a list of people who were often ignored, or even ridiculed by their peers. Their names include Cantor, Peano, Koch, Sierpinski, Hausdorff, Julia, Fatou and Richardson; their work involved *Dimension, Iteration Theory, Self-Similarity,* and *Measurement.* The work of each would "far transcend its intended scope" and their often unconventional approaches to the "unexplained" would later spark a surprising synthesis of their work [Mand82].

Dimension Crisis 1875-1925

History is filled with examples of struggles to hold on to old definitions of the world rather than face the new and unexplainable. Mathematics of the 19th century had become no exception. Euclidean Geometry and Newtonian Physics were deeply rooted traditions, and if a phenomenon failed to fit within the boundaries of these theories, it was immediately labeled an exception or even a monster. (See *Figure 1 & 2.*) Toward the end of the century, these "exceptions" were becoming so common that by 1875, they could no longer be ignored. In that year mathematician DuBois Reymond shocked the mathematical community by publicizing what Karl Weierstrass had presented three years earlier. This "object" was a mathematical function that, unlike other functions of the day, was continuous though nowhere differentiable. In other words, the graph, or pictorial representation of this function was without breaks, and yet nowhere could one attach just one tangent line. (See *Figure 3 & 4.*)

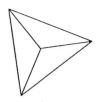 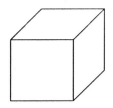 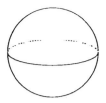

Figure 1 Euclidean Geometry

The two thousand year formalized study of Euclidean Geometry makes it familiar to all of us. From architecture to modern art, it is clear that euclidean shapes dominate our human made environment. This continues today, and whether for a kindergarten child learning her or his first shapes or for a high school geometry student, tradition has dictated that teachers choose examples from a library akin to those figures above. Now what of those below?!

Figure 2 Newtonian vs. Non-Newtonian Physics

If we assume that we are able to completely describe the initial conditions of a physical situation, then using Newtonian physics we should be able to completely describe its outcomes. Though this science has completely revolutionized the world we live in, scientists now realize that a different approach is required for those cases where it is impossible to precisely describe all beginning criteria. The figure on the left is a classic beginning physics problem. Given all the beginning criteria (forces on the object of mass m, length of the ramp etc.), what will happen? The figure on the right, however, is a different story.

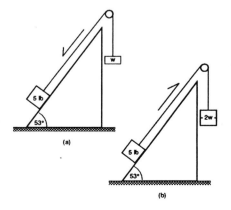 

HENON MAP

Figure 3 Weierstrass Function

In 1872 the Weierstrass Function was exceptional; although it was continuous, it was not differentiable. Repeated magnifications show that what may appear at some distance to be a smooth section of the graph of the function actually contains an infinite series of peaks and bumps.

Figure 4 Tangent lines

A tangent line is a line that touches a curve at just one point. For the figure on the left, there is only one line that touches the curve at point A. However, for the curve on the right, there are several, and in fact an infinite number of lines that touch the curve at that point.

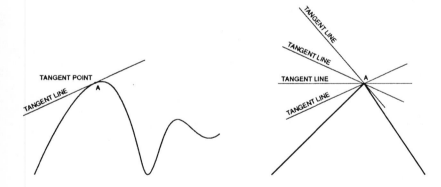

To use a non-mathematical analogy, although the graph of the Weierstrass Function continues without holes or interruptions, the smallest airplane trying to use the function as a runway would never be able to land. Repeated magnifications show that what may appear at some distance to be a smooth section of the "runway" actually contains an infinite series of peaks and bumps. With regard to this function DuBois Reymond wrote, "These functions seem to hide many puzzles, as far as I am concerned, and I cannot rid the thought that they will lead to the limit of our intellect" [Mand82].

The problem with this function was that it contradicted the basic premise of calculus. Calculus is built around the idea that "A continuous series is differentiable because it [normally] can be split up into an infinite number of absolutely smooth straight lines." However, "a non-differentiable continuous series cannot be resolved. Every attempt results in still more roughness" [Burr81].

In order to successfully confront a "monster" such as the Weierstrass function, mathematicians soon realized that they would have to reassess their understanding of the term *dimension*. (See *Figure 5*.) Dimension was, and is, discussed in terms of a set of coordinates; one coordinate describes a point on a *one*-dimensional object, two coordinates describes a point on a *two*-dimensional object, and so on. However, this simple classification was inadequate for a proper understanding of the "pathological" objects that Reymond had come across [Mand82]. These complicated figures demanded a reevaluation of the traditional concept of dimension, and as a result, mathematicians reluctantly acknowledged a dimension crisis in their field that spanned the next 50 years.

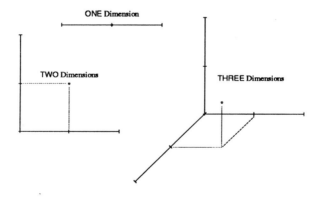

Figure 5 Dimension

Dimension was, and still is discussed in terms of a number of coordinates. One coordinate describes a point on a line, which is one-dimensional. Two coordinates describes a point in the plane, which is two-dimensional, and three coordinates designates a point in space, which is three-dimensional.

The Pioneers

A rigorous examination of dimension ran concurrently with more curious examples. Of these, some of the most disturbing came from Georg Cantor, Guiseppe Peano, and Helge von Koch. This list of pioneers begins with Cantor, whose contributions surfaced two years after the unveiling of the Weierstrass function. By 1883 he had produced the Cantor set, the first of several unprecedented objects. (See *Figure 6*.)

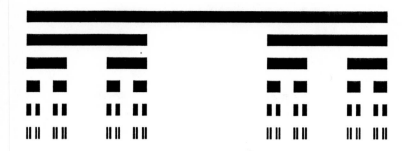

Figure 6 Cantor Set

Though the investigation of this set now pervades any introductory class on fractals, it was avoided by Cantor's peers. This set is constructed by successively removing more and more sections of the unit interval. The process is repeated ad infinitum and the infinite number of points that remain constitute the Cantor set.

This set is constructed by beginning with the piece of the number line from 0 to 1 (known as the unit interval), removing the middle third of it, and then applying the process again to the remaining two pieces. This leaves four segments, from which the middle thirds are again removed, leaving eight strips. This process is repeated over and over, and theoretically forever. The infinite number of points that remain constitute the Cantor set. Though Cantor's work was shunned by his peers throughout his life and for many years after his death, he is now often considered the founder of modern mathematics. As late as 1962 Cantor's set was considered "at least as monstrous" as its counterparts, the Peano and Koch curves [Mand82]. (See *Figures 7 & 8.*)

Like Cantor, Peano's work was also considered to be of a foreboding nature. The Peano curve, like Cantor's set, is constructed by applying the same process to successive generations. Normally, Peano's mathematical formula would be thought to produce the picture of a one-dimensional curve. But after several generations, the Peano curve appears to fill the plane. To mathematicians this was unsettling because an object that fills the plane is by definition two-dimensional. For this reason these plane-filling curves of the 1890's sparked descriptions such as non-intuitive and extravagant. To one mathematician, their consequences were thought to do no less than "cause all basic mathematical concepts to lose their meaning" [Mand82]. Another aberration had surfaced, and to the majority of the mathematical community, this meant that "everything was in ruins" [Mand82].

Figure 7 Peano Curves

Like the Cantor set, the Peano curve is constructed as a result of applying the same process to successive generations. Though curves are considered to be one-dimensional, it is clear that after an infinite number of generations the Peano curve will completely fill the plane. But this was unsettling: an object that fills the plane is two-dimensional. In short, Peano's work suggested that a one-dimensional curve could fill a two-dimensional plane! For this reason these curves were labeled "counter-intuitive" and "extravagant" by Peano's peers [Mand82].

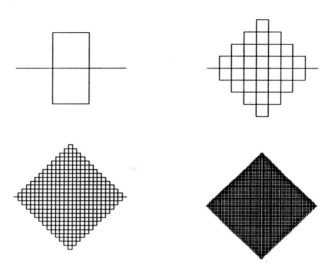

About a decade later Helge von Koch introduced the first of what we now call the Koch generated curves. This curve was unsettling because it was known as nonrectifiable; that is, like the Weierstrass function, one could neither land a plane on it nor measure how long the "runway" was. Mathematicians did not have the appropriate theoretical tools, and without them, because repeated magnifications revealed more and more bumps, attempting to measure the "length" of the "runway" did not make sense. Using traditional methods to measure these strange curves consistently yielded undesirable answers. Although these curves were comfortably contained in a finite area, their length according to standard methods, was infinity! Similar to the Peano curves, these objects seemed to evade even the most basic of mathematical manipulations. The study of such deviant sets was questioned by conservative mathematicians. Non-differentiable curves such as the Koch curves were considered by almost every account "counter-intuitive, monstrous, pathological, or even psychopathic" [Mand82].

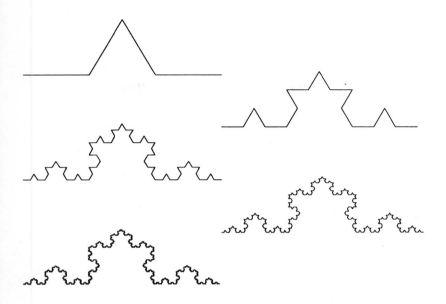

Figure 8 Koch Curves

Like the Weierstrass function, the Koch curves were yet another example of the continuous though nowhere differentiable functions. Worse yet, they were nonrectifiable; that is, mathematicians had difficulty measuring their length. Using standard methods of measurement, after an infinite number of generations, the length of the curve above was found to be infinity (an infinite number of generations will yield an infinite level of detail; shown above are just 5 generations). But how can a curve of infinite length occupy and be contained within a finite space or area?! Though we are limited both by our eyes and printing capabilities, if one were to draw a curve that NEVER ended, how does it fit in a finite interval (which it theoretically does) without overflowing?! Because they could not answer this question, mathematicians felt that the functions associated with these pictures were "counter-intuitive."

By the turn of the 19th century it was clear that mathematicians were suffering a crisis of intuition. On the one side was the very large school of thought that believed that intuition did not allow for these continuous functions without derivatives (or "non-landable" runways). On the other side was a small and all but silenced minority that felt just the opposite; not only were these functions intuitively accessible, but more naturally so.

Time would vindicate the silenced; the true nature behind these Cantor sets, Peano curves and Koch curves would show that they were not nature's exception, but very often its rule. Now, nearly a century later they mark the beginning of the history of Fractal Geometry, and despite any intentions of their founders they are considered "classical fractals."

The key to resolving the dimension crisis caused by these exceptional cases came in addressing the fears of mathematicians head on. The results of measuring these curves and sets did not "make sense" (because they were either 0 or infinity) *as long as the standard definition of dimension was applied.* However, if the concept of dimension could be expanded, these measurements would become meaningful and the so-called monsters would be tamed.

The problems presented by these objects were eventually addressed by a philosopher-turned-mathematician named Felix Hausdorff. In 1919 Hausdorff defined a new kind of dimension for standard objects. A.S. Besicovitch later generalized Hausdorff's work, and now the Hausdorff-Besicovitch dimension is one that is associated with shapes in such a way that allows for *non-integer* dimensions [Mand82]. The dimension of an object was no longer restricted to values such as 1, 2, or 3, but could actually lie somewhere *in between.* This did not eliminate the traditional concept of dimension, but rather offered an alternative approach. That is, with Hausdorff's approach a set could theoretically have fractional dimensions such as 1.4, or 1.8, or 2.7!

Though it would take nearly a century to refine, the dimensions of the curious examples put forth by Cantor, Peano, and Koch, can now be clearly explained in terms of this quantity (See *Chapter 3*).

Julia, Fatou, and Iteration Theory

While Hausdorff was tying up the loose ends of the Dimension Crisis, two French mathematicians, Gaston Julia and Pierre Fatou, were working on what is called *iteration theory.* Iteration theory is built around the concept of a continuous loop. A convenient modern tool for explaining iteration theory is the scientific calculator. If we were to push the button "3" and then "x^2" the calculator would compute 3x3=9. Hitting the "x^2" key again will compute 9x9 or 81. Repeatedly hitting the function key "x^2" repeatedly sends the result of the previous computation back into the function "x^2"

Similarly, in iteration theory, a value is put into a mathematical process called a function, or rule. The function then churns out the next value; this in turn is fed back in, and the process is repeated. (See *Figure 9.*) To introduce the vocabulary, using the *initial value* 3, one *iterates* the *function* [f(x) = x2], and we say the result of the *first iteration* is 9 and the result of the *second iteration* is 81. The list of numbers 3, 9, 81… is called the *orbit* of the initial value 3. For this particular orbit, the values get greater and greater, and actually tend toward infinity. This highlights an important question: will an orbit (which varies for different functions, as well as for different initial values) tend towards infinity, or will it tend towards a finite number and thus stay *bounded?* Using complex numbers* mathematicians such as Julia and Fatou were interested in those patterns and trends that occur after many such repetitions.

Though they worked independently of one another, Julia and Fatou both focused on iterating rational functions of a complex variable. Most importantly, they investigated whether the results for a certain initial value got continually greater, or instead settled down to some finite number. The theme of this work was that iterating even the simplest of these mathematical functions could produce very complicated results [LaBr87].

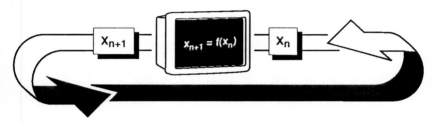

Figure 9 Iteration

Central to Julia and Fatou's theories was the concept of iteration. Iteration may be thought of as a continuous loop. A value is put into a function, the function churns out the next term; this in turn is fed back in, and the process is repeated. For example one can "feed" the number "3" into a rule such as f(x) = x². The result is (3)² = 9. Now if one feeds the 9 back into the function the result is (9)² = 81. 81 is then fed back in, the result is 6561, and so the process continues.

The theories set up by Julia and Fatou laid the foundation for generating what we now call Julia sets. (See *Figure 10.*) Unfortunately for Julia and Fatou, these complex "objects" remained statistical because they could not be manually drawn. The list of values that resulted from even the simplest of their investigations was so vast that it was visually indecipherable. Though they had a keen intuitive sense of the beauty of their results, this beauty could not, and would not be seen until almost 60 years later with the advent of the computer.

---

* (numbers of the form a+bi, a, b real, i = √-1)

Figure 10 Julia Sets

With the help of the computer, the beauty and complexity of these Julia sets is now apparent. For Julia and Fatou, however, the list of values that resulted from even the simplest of their investigations was visually indecipherable. The computer-generated figures above represent the boundaries of sets whose members are those initial values for which the orbits remain bounded.

Self-Similarity

While Julia and Fatou were working on iteration theory, an English scientist by the name of Lewis Richardson was studying two other areas: the behavior of turbulence and a new way to assess the length of coastlines. An example of turbulence is the departure from a smooth flow that occurs when a fluid hits an obstruction or exceeds a certain speed (picture in your mind a free flowing stream that suddenly hits a rock). Extremely complicated and not completely understood even now, turbulence had long been the untouchable subject of physics. In 1926, Richardson made a qualitative observation concerning this subject: he noted that "over a wide range of scales turbulence is decomposable into self-similar eddies" [Mand82]. He illustrated this self-similarity with a parody of Jonathan Swift's, inspired by the great mathematician G.W. Leibniz.

> *So Nat'ralists observe, a Flea*
> *Hath Smaller Fleas that on him prey.*
> *And these have smaller Fleas to bit 'em*
> *And so proceed ad infinitum.*

Jonathan Swift 1733

Concerning his findings in turbulence, Richardson explained:

Big whorls have little whorls,
Which feed on their velocity;
And little whorls have lesser whorls
And so on to viscosity
(in the molecular sense)

Lewis Richardson 1926

Leibniz (1646-1716) had pointed out the self-similarity of a line already in the late 17th century. He noticed that a line is made up of small copies of itself. That is, it is similar on smaller scales. He states "The straight line is a curve, any part of which is similar to the whole, and it alone has this property, not only among curves but among sets" [Mand82]. However, after this initial observation the geometrical discussion of self-similarity within the mathematical community seemed to cease. Although this property of self-similarity, also known as *scaling*, was discovered independently in a wide variety of fields, "the actual study of scaling...has spurned geometry, and has overwhelmingly concentrated on analytical considerations" [Mand78].

Richardson's observations, though presented in a playful rhyme, were extremely important. He recognized that *patterns exist within the geometric structure of a physical system that are similar on finer scales*. In the case of turbulence, there is a visible pattern of whorls within whorls within whorls. Turbulent movement exhibits self-similarity.

How Long is a Coastline?

How long are the coasts of Britain, Japan, Cuba? The boundary between Brazil and Peru, or India and Pakistan? The discrepancies in the measurement of common borders between countries are surprisingly commonplace. Lewis Richardson explained the disparity of these measurements when he pointed out that the measured length of a coastline will depend on the size of the measuring stick used. For example, if children were to use a standard map of California to measure the coast from San Francisco to Los Angeles, they might measure some 450 miles. If, however, they used a map that showed more bays and inlets, this measurement might increase to 500 miles or more.

Why? A coastline measured with say, a yardstick would have a shorter measured length than the same coastline measured with a 1 foot ruler. When using the measuring stick of one yard, despite the fact that some indentations would be missed, a length for the coastline could be determined. However, a more accurate length would be determined by using the measuring stick of one foot. The one-foot ruler allows for the measurement of more inlets in the coastline, thereby increasing the total distance measured, and yielding a longer length.

Thus, as the measuring stick gets shorter and shorter, the measured length gets longer and longer. How long is the coastline? If we assume that our measuring stick can shrink indefinitely, then the coastline is infinitely long. (See *Figure 11*.)

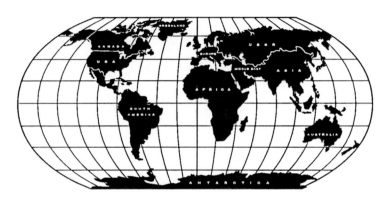

Figure 11 The Coastline Question

Using this reasoning, as the map becomes more and more detailed, in fact, infinitely more detailed, it becomes impossible to obtain a true length for the coastline. This is the same problem mathematicians came across when attempting to measure the lengths of the Koch curves, the Weierstrass function, and our "runways". Each of their lengths was infinity as well, and in mathematical terms these examples were called *nonrectifiable*.

Richardson's conclusion was that the calculated length of a coastline can change depending on the unit of measurement used. To avoid a common misunderstanding, it is important to stress that physically the length of the coastline does not change; but our calculated measurement of it does. This is equally disturbing though. At first it may be difficult to understand why the coastline question and its implications would puzzle scientists. But imagine the complications if measuring the volume of a quantity of water changed depending on the size of the container used to measure it.

Similar to the problems encountered during the dimension crisis, previously accepted methods were again inadequate for addressing the complexity of measuring these objects. Like the Koch curves, using standard methods of measurement to calculate the length of a coastline yielded an infinite length. The coastline was yet another non-rectifiable example.

Richardson attacked this problem head on. Using a *polygon*, or "many-sided" figure, as a model for a country's boundaries, Richardson concluded, "the polygon representing the coastline has a number of sides inversely proportional to the step length (measuring stick length) raised to a certain power, d, which depends only on the contour (of the coastline) to be traversed" [Mand78]. Translated into mathematics this relationship is described by the equation $N = k/l^d$, where N is the number of sides of the polygon, l is the length of the measuring stick, and k is some constant. In layperson's terms, the length of the coastline depends on how small the measuring stick is: the smaller the stick, the greater the measured length.

Mathematical equations aside, it is important to note that Richardson had devised a new means of measuring what in the strictest sense had been immeasurable. Theoretically, like the curves of the dimension crisis, the length of the coastline could be infinite. However, Richardson's equation, dependent on the "certain power d," gave a much more meaningful way to assess this length. The value denoted by "d" was to become very important in relating measurement to the dimension of both the coastline and the seemingly obscure products of the dimension crisis. Why? Long after Richardson's death, a young mathematician was to come across the scientist's work and be the first to explain how the "d" obtained by measuring the length of this nonrectifiable curve was in fact the Hausdorff-Besicovitch dimension of the coastline [Mand82].

Waclaw Sierpinski belonged to a group of mathematicians who, against the advice of their contemporaries, worked at constructing a new range of mathematical figures. These figures, though relatively simple to construct, seemed to defy conventional means of measurement. Something as fundamental as measuring the length, area, or volume of these sets perplexed some of the best mathematicians of the day. As a result, both the mathematicians and their projects were described with a host of less than charitable adjectives. By 1916 Sierpinski had contributed to this "gallery of monsters" by investigating the construction of a figure that we now call the Sierpinski Triangle. Like the work of Georg Cantor, Guiseppe Peano, and H. Von Koch, his recursive algorithms, or "recipes of instructions," furnish quite interesting results.

Just as a Cantor set can be constructed by successively removing middle thirds of a line segment, a Sierpinski triangle can be made by successively cutting out triangles. Using the midpoints of each side as a vertex, a smaller inverted triangle is drawn in the middle of the figure and then removed. This process is then applied to the remaining triangles, and is repeated without end.

One of the interesting characteristics of the Sierpinski triangle is revealed by investigating its area. The "uncut" triangle in the figure below has an area of 2 in². However, when we cut out the first triangle we remove with it 1/4 of the original area, leaving only 1.5 in². The next step leaves an area 1.12 in², the next with 0.84 in² and so on. But this is repeated forever. What is the resulting area? After an infinite number of generations— the area is 0!

At the same time we can look at the length of the boundary of the "holes" we leave when we remove successive triangles. As the number of "cut out" triangles increases to infinity, so will the perimeter of these holes. *In this way, the length associated with the triangle is infinite and yet the area associated with it is 0.* The triangle, like other "classical fractals," yields unpleasant results when investigated with standard methods of measurement. The one-dimensional "size" of the figure (what we normally call length) is infinite, and yet its two-dimensional "size" (what we call area) is zero.

Thanks to a few adventurous mathematicians who broadened the concept of dimension, we can now say that the triangle hovers *between* the first and second dimension. To determine where, mathematicians now calculate its "fractal dimension." The fractal dimension of the Sierpinski Triangle is 1.584, a little more than half way in between the first and second dimensions. This 1.584th dimension is the only dimension for which the "size" of this object makes sense, (not 0 or infinity).

 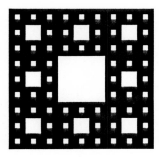

Sierpinski Triangle & Carpet

The fractal dimension of the Sierpinski triangle is approximately 1.584, which means it lies a little more than half way in between the first and second dimensions. How would a figure with fractal dimension of say, 1.892 differ? Would you expect it to have "more" or "less" holes cut out? Compare the Sierpinski triangle with that of the Sierpinski carpet, which does have fractal dimension of approximately 1.892. Do the pictures agree with your expectations?

It should be pointed out that theoretically, each of the figures above has an infinite number of holes "cut out." Yet their visual approximations lead us to think that the Sierpinski triangle, as opposed to the carpet, has somehow more holes. The length associated with the triangle would seem to be "more infinite" than that associated with the carpet; conversely, although each has 0 area, the triangle seems to have less 0 area. In a conventional sense such musings are absurd, (hence the adjectives "monstrous," and "psychopathic"). However, enter the laws of this fractal dimension, and suddenly there are numbers associated with these sets that confirm our observations.

Sierpinski Pyramid

The picture above is generated in the same manner but the initial set is a pyramid. Where would the dimension of this set fall? Between 0 and 1, 1 and 2, or 2 and 3? Can you guess what its fractal dimension might be?

"Mandelbrot's work in Fractal Geometry gives insight into the complex shape and structure of the natural world: the turbulence of liquids, the symmetry of living forms, the branching of crystals or rivers, the fluctuation of radio static, and the stock market."

[MInt85]

What were the connections, if any, between *Dimension, Iteration Theory, Self-Similarity,* and *Measurement?* Among the small group of mathematicians investigating these diverse areas, some very important groundwork was being laid for modern mathematics. But who would have suspected it? Had it not been for the French mathematician Benoit Mandelbrot, we might not know. He took these seemingly unrelated ideas, identified their similarities, and using his experience and intuition, made his mark in mathematical history as the person credited with establishing Fractal Geometry. For many, the relationships between these diverse fields were so unlikely that when Mandelbrot was ready to unveil his findings, he was often rebuked. His work was self-described as "that which is published only when the censors nod"; and according to Mandelbrot, the censors were none too kind. More than once, fellow mathematicians or editors recommended that his work remain unseen "until the world was ready for it" [Mand82].

Mandelbrot's introduction to the ideas of dimension, iteration theory, self-similarity, and measurement began in the mid-forties and continued until the early sixties. When Mandelbrot was 20 years old his uncle insisted that he read Julia's papers on iteration theory. This had a decided impact on his career and he later became a student under Julia. During his education he was also exposed to the expansion of dimension theory that followed the discovery of the Weierstrass function. This included Cantor's work, though at the time it was considered to be of little practical use.

It was outside the traditional mathematical arena that Mandelbrot began to recognize the concepts of self-similarity. Lewis Richardson provided confirmation of these self-similar patterns, as well as a discussion of measurement in the coastline question. From the mid 60's on, Mandelbrot's work on a link between the four areas can best be discussed in terms of dimension and self-similarity.

> *"The oscillation between different fields of science and mathematics has been a constant, a hallmark of his entire career."*

[LaBr87]

Mandelbrot obtained his Ph.D. in Mathematics in 1952 and from there went on to become what is best described as a "Mathematical Jack of all Trades" [Glei88]. In the early 60's, while analyzing error distributions for Bell Laboratories, he noticed a pattern in the frequency of telephone line errors. Though at first glance the errors seemed to occur randomly, Mandelbrot found otherwise. His data showed that there would be a clump of errors and then none for a while, followed by a clump of errors and then none, and so on.

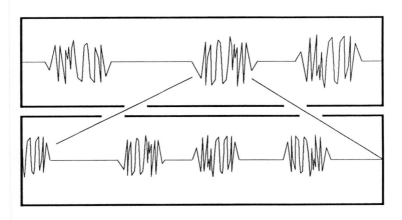

Figure 12 Experimental Self-Similarity

However, when he examined one of these clumps more closely, he found within it a small-scale repetition of the larger pattern. (See *Figure 12.*) To Mandelbrot's surprise these patterns were also evident in other areas of research. His charts of fluctuations in the level of the Nile river, and fluctuations in commodity prices also exhibited this repetitive self-similarity. In the discussion of turbulence, Richardson had pointed out a similar phenomena of "whorls within whorls." Motivated by these observations, Mandelbrot began to look more intuitively at the connections between these apparently unrelated situations.

A *model* is a mathematical tool that simulates or approximates a real life situation. Though they are usually highly simplified versions of the systems they attempt to duplicate, models are extremely valuable. With the aid of computers, they allow physicists, economists, biologists and mathematicians to simulate experiments that might otherwise take months or years to observe.

Much, if not all, of the goal of applied mathematics is to find models of physical systems. Mandelbrot had found evidence, as had Richardson, of self-similarity in physical systems. Telephone errors, for example, were distributed in a way that exhibited self-similar patterns over time. But what of a mathematical model for this distribution that would also exhibit self-similarity?

As testimony to his far reaching creativity, Mandelbrot made a remarkable connection between the scaling effect and a mathematical entity that had once seemed as far removed from reality as possible. While the rest of the mathematical community had decided Cantor's work was useless, Mandelbrot realized that the Cantor set discovered in 1883 could serve as a rough model for telephone error distributions. As he says in *The Fractal Geometry of Nature:*

> *"We construct the set of errors by starting with a straight line, namely the time axis, then cutting out shorter and shorter error free gaps. This procedure may be unfamiliar in natural science, but pure mathematics has used it at least since Cantor. As the analysis is made three times more accurate, it reveals that the original burst is intermittent."*

[Mand82]

The construction of the Cantor set matched the "clump-void" pattern of telephone error distribution. Like the clump of errors, each bar of the Cantor set is also broken into thirds, and the middle third (the error-free section) is left out. As with any good model, this allowed scientists to investigate telephone error distributions by studying the particulars of the Cantor set.

With these observations, Mandelbrot established a connection between a physical system and the-all-but discarded geometry of the dimension crisis. Furthermore, he was able to shed light on systems thought to be entirely random. Reluctance to accept Cantor, however, was widespread and though Mandelbrot had first published 10 years previously, his newest work was rejected until he had completely disguised any signs of Cantor's influence [Mand82].

Mandelbrot's sensitivity to these self-similar patterns would later enable him to recognize yet another example. With the onset of the computer age he was able to determine the existence of self-similarity in those patterns displayed by the results of the Julia-Fatou iteration process. When drawn with the aid of computer graphics, Mandelbrot found that though these shapes were not always exactly self-similar, they did exhibit a certain degree of repetition on finer scales.

By using the Cantor set as a mathematical model of a self-similar system, Mandelbrot had succeeded in revealing the relevance of an isolated and supposedly useless shape. Furthermore, though the scientific community had not asked, in just this one example Mandelbrot was able to show the connection between a seemingly random physical system, self-similarity, iteration theory, and a product of the dimension crisis.

Early Observations: Dimension

Mandelbrot studied both boundaries and curves, and with his keen eye the correlations between the jagged coastline, the pointed Koch curve, and the infinitely bumpy Weierstrass function became apparent. All were characterized by their irregularity, and all had presented difficulties when approached with standard methods of measurement. Because we normally use length, area, or volume as an essential part of an object's description, these outlandish sets were more than just perplexing. They were in conventional ways immeasurable, and therefore mathematically, they were indescribable.

Historically, the Weierstrass function required that mathematicians reevaluate the usefulness of the conventional idea of dimension; and the Koch curve necessitated a new approach to measuring length. When a problem similar to these was posed and addressed by Richardson, Mandelbrot recognized that the answer to taming these shapes lay in the generalization and extension of Richardson's findings.

Richardson had unwittingly begun unraveling the mysteries of these curves when he detailed a sensible way to measure the length of a coastline. Mandelbrot found that work by chance, many years after it was written. As he examined the work more closely, he realized that unknown to Richardson, the method of measurement was dependent on the Hausdorff-Besicovitch dimension of the coastline. Though the idea had been hidden for decades in this obscure paper on coastlines, Mandelbrot discovered that this new connection to Hausdorff's dimension would allow Richardson's techniques to be applied to these enigmatic shapes.

The largest step in taming these nonrectifiable curves lay in Mandelbrot's diagnosis that their measurement made sense only if this alternative idea of dimension was used. This application of Richardson's approach to the Koch curves highlighted theoretical as well as physical evidence that solidified the existence of shapes that were essentially between dimensions. The theory now allowed for *fractional* dimensions, and in fact, Mandelbrot was able to show that the Hausdorff-Besicovitch dimension of the coastline of Britain was 1.26. The calculated dimension of the Koch snowflake was shown to be 1.2618 [Mand82]. (See *Figure 13.*) (Also see *Chapter 3: Calculating Dimensions.*)

This new dimension would eventually become the central tool for measurement of these unique shapes. Like the length, area, and volume of Euclidean shapes, fractal dimension is now indispensable in the descriptions of fractals.

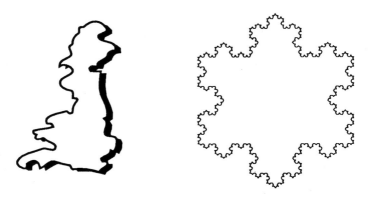

Figure 13 Britain and the Koch Snowflake

Mandelbrot was able to show that the Hausdorff-Besicovitch dimension of the coastline of Britain is 1.26. The calculated dimension of the Koch snowflake was shown to be 1.2618.

On the Outskirts of Traditional Mathematics

"Fractal Geometry is a new example of an historical anomaly."

[Mand82]

As a mathematician, Mandelbrot first published in 1951. Twenty five years later, with the release of *Les Objets Fractals*, he succeeded in drawing together the four areas of *Dimension, Iteration, Self-Similarity* and *Measurement.* The techniques with which he connected these ideas show the mathematically rare reliance on description as a precursor to theories. Though he had been warned by his uncle that "mature mathematicians do not use visual images" [LaBr87], he demonstrated an even more unusual trust in vision to form conjectures and recognize patterns. The success with which his conjectures evolved into Fractal Geometry suggest that the use of images is a welcome and essential tool for understanding this new branch of mathematics.

Even today, however, mathematicians reject this opinion. Mandelbrot's work was not always considered legitimate mathematics, and this may explain why, between the years of 1951 and 1975, he had largely been working alone. "It is rare," Mandelbrot observes, "that you get a new field of science born in the absence of competition." But, according to Mandelbrot, that is what happened. Despite the lack of competition or collaboration, he was able to establish theories that would finally address the geometry of nature. Who would have expected it? Even the creators of some of the unusual examples had failed to see their link to nature's geometry.

Though there were abundant counter-examples, the belief continued to hold that the chaos found in nature was the exception. F.J. Dyson summarizes:

> *"These new structures were regarded…as 'pathological' as a 'gallery of monsters,' kin to the cubist painting and atonal music that were upsetting established standards of taste in the arts at about the same time. The mathematicians who created the monsters regarded them as important in showing that the world of pure mathematics contains a richness of possibilities going far beyond the simple structures that they saw in nature. Twentieth century mathematics flowered in the belief that it had transcended completely the limitations imposed by its natural origins."*

> *"Now as Mandelbrot points out, Nature has played a joke on the mathematicians. The 19th century mathematicians may have been lacking in imagination, but Nature was not. The same pathological structures that the mathematicians invented to break loose from the 19th century naturalists, turn out to be inherent in familiar objects all around us."*

[Mand82]

Mandelbrot did not build the theory of fractals by himself. However, one can easily say that without him, Fractal Geometry would not yet be. With the first publication of *The Fractal Geometry of Nature* in 1977, Mandelbrot solidified the existence of a new branch of mathematics.

Even so, this was only the beginning. Three years later, Mandelbrot's search for a categorization of Julia sets led to the discovery of the set that now bears his name. This Mandelbrot set contains, literally, an infinite number of fractals. Its discovery can be likened to discovering a galaxy with an infinite number of stars, only to find that there exists an infinite number of galaxies. (See *Figure 14.*)

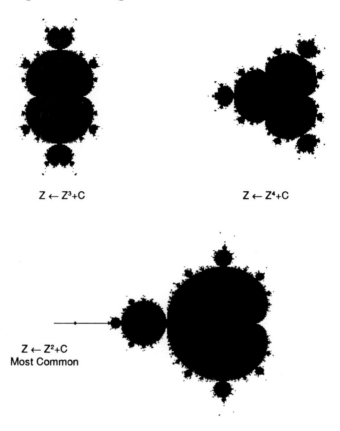

$$Z \leftarrow Z^3 + C \qquad\qquad Z \leftarrow Z^4 + C$$

$$Z \leftarrow Z^2 + C$$
Most Common

Figure 14 Mandelbrot Sets

Mandelbrot's search for a categorization of connected vs. non-connected Julia sets led to the discovery of the set above ($z \leftarrow z^2 + 2$). Certainly one of the most beautiful mathematical objects, this set contains literally an infinite number of fractals. Its discovery can be likened to discovering a galaxy with an infinite number of stars, only to find that there are an infinite number of galaxies. Though recent literature seems to focus on only one, there are as many Mandelbrot sets as there are mathematical functions.

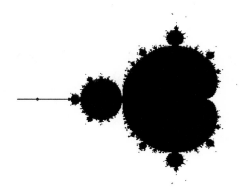

The above set was first seen in 1980 when it was generated using a computer at Harvard University. It is without a doubt one of the most beautiful and complicated mathematical objects ever discovered. But what is it? To some it serves as an infinite table of contents associated with Julia sets [Peit86], to others it is just "a snowman with a bad case of warts." But, as they say, every picture tells a story and the Mandelbrot set is no exception.

A useful way of understanding the story behind the set is to look at how and why it was constructed. Mandelbrot, in his search for a way to categorize results of the Julia-Fatou iteration process, found two types of Julia sets: those that were connected, and those that were totally disconnected. Using the function x^2+c, he attempted to get a snapshot of all those "c" values (points in the plane) for which the associated Julia sets were connected.

Imagine it to be a hot sunny day and you have been designated as team score-keeper at a baseball game. Part of your job is to keep track of each player as they go up to bat. Only there are about 150,000 batters on your team. To keep things organized you use a grid. It looks like a sheet of graph paper, with each box representing a player. To simplify your task further you will keep track of only one thing on your team roster: whether or not the player at bat can hit a ball out of the ballpark. If the player hits the ball out, leave the square blank; if the player does not hit the ball out of the ballpark, color it black. Simple.

Now what would you expect your grid to look like? A random scattering of black and white perhaps, or based on your knowledge of the team, perhaps a slight pattern?

Batter up. First player is Homer. He swings. It's...it's gone! This one is out of the ballpark. On the grid, the square for this player is definitely left blank.

Next batter. Ok, here we have Solid Sue. A player known for continuity. A real solid hitter. Here's the first hit! It's in the park, looks bounded; a nice base hit down the line. That one stayed inside. Color that square!

Next player, Questionable Cory. A player who has good days and bad. Have to watch this one carefully. You think, "Here's the base hitter all the way," and then on that last pitch...Boom. It's outta here. Just depends, you know. Seems to be the more pitches he gets, the more likely it is he'll hit it out. Here's that last pitch, looks close...No, just shy. One more chance might have made the difference. Maybe next game. In the mean time color that square.

_____

(150,000 batters later...) Whew, what a game. Thought we'd never finish that one. Well how'd we do? Let's take a look at that roster. See if we can't make some sense of all this.

Uhhhhhhh. Hey uh, Charlie, I don't think you're gonna believe this one.

That's unbelievable!

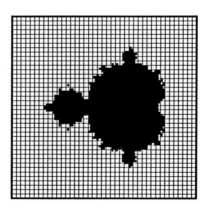

And it was unbelievable. Imagine Mandelbrot in 1980 as he waited to see what his plot of connected vs. non-connected Julia sets would be. This Mandelbrot set, though it had a very straightforward construction, resulted in a figure that is intricate, and at times complicated beyond belief.

One writer called the program that generates the Mandelbrot set, "deceptively simple." But it *is* simple. Less than three pages of computer code are enough to test the values. Using the equation $x^2 + c$, let x be zero, throw in a player (c), let the computer do the iterations, check to see if they're in or out of the ballpark, color (or don't color) the associated square, and then go on to the next player (next c). The computer screen is essentially the team roster; one pixel (or square) per player.

Because of time constraints, one can arbitrarily choose, say 32 iterations, and if after these 32 iterations, or "pitches," the "player" still hasn't "hit it out" (orbit hasn't gone off towards infinity), tell the computer to color the corresponding square on the screen black. The assumption is that if they don't hit it out of the ball park after these 32 pitches, they never will. This isn't always the case though, and crisper pictures can be obtained by allowing a player a longer time (more pitches and so more chances) to hit it out. This catches "late bloomers," such as "Questionable Cory" and therefore the resulting Mandelbrot set is more accurate.

Again, the Mandelbrot set is a picture of the values whose associated Julia sets are connected. If the Julia set is connected, then the orbit obtained when the initial value "c" is iterated will eventually settle down "inside the ball park." The instructions are then to color the associated square black. Solid Sue, for example would have a connected Julia set such as the one in *Figure A*.

If, on the other hand the associated Julia set is disconnected, then the orbit of the initial value c, soars off, "out of the ballpark," to infinity. The associated square of the "team roster" is then left blank. Homer, therefore, would have a disconnected Julia Set such as the one in *Figure B*.

Figure A
Solid Sue's Julia Set

Figure B
Homer's Julia Set

FRACTALS

"For reasons we don't really understand nature appears to organize herself according to Fractal Geometry."

[Kozl86]

GEOMETRY INSIDE AND OUT

Look up.
Look around the room or building you are sitting in.
What kinds of geometrical shapes do you see?
Squares, triangles? Straight lines, right angles?
Now–
Look Outside.
What kinds of shapes do you see?
Certainly they are different from the smooth,
well-ordered objects inside.
What of the trees, the clouds, a mountain, or a fern?
Though not Euclidean, what of their shape, order and structure?
Can they be considered geometrical?!

Until a few years ago most would have answered "no". Today, however, Fractal Geometry has extended the investigation of form and structure to those objects found in nature. As a result, the study of geometry is no longer confined to the classroom, but reaches beyond the realm of human made structures to the landscape outside.

Previously, much of mathematics has seemed distant to the layperson and unrelated to the world around us. Now, with the expansion of geometry to include fractals, students of all ages can study a geometry for "outside" as well as "inside." Richard Voss, whose extraordinary fractal landscapes have accompanied Mandelbrot's books, says:

"Fractals, with their complicated regularity, are closer to the shapes nature makes than are the simple regular geometric figures we have been used to; circles, triangles, rectangles, etc. The geometric figures are characteristic of human artifacts. Fractals are characteristic of nature."

[Thom87]

Once introduced to the world of fractals, the intricate, lacy patterns of natural objects seem illuminated as never before. Whereas the straight line dominates the geometry of the "inside," the branching patterns of fractals seems to appear again and again in nature. From the web of arteries and veins to the structure of river networks, what we have always considered unstructured suddenly displays an underlying order.

The contrast to the smooth euclidean diet that geometry traditionally offers seems at first harsh, but in fact, for many new explorers of Fractal Geometry, these objects provide a missing link in mathematics education. Not only do fractals formalize the study of nature's geometry, but they begin to chip away at the belief that math is accessible only to the few, and enjoyed by still fewer. As one teacher points out, "Fractal Geometry can help overcome the intimidation stemming from the myth that mathematics is a subject built from logic upon a firm, well-defined base which, by its very construction, does not allow for contradictions or incongruities" [Whit89].

A Visual Introduction

Introducing fractals without touring their images would be like introducing a beautiful symphony by displaying only the score.

Mathematicians believe that using a visual image to make a conjecture can be dangerous. For this reason the short history of Fractal Geometry is punctuated by rejection from a substantial portion of the mathematical community. The problem does not always lie with the image itself, but with the interpretation of the image. True, those conjectures made by visual observation should ultimately find support in proofs. However, the amount of research sparked by the visual pondering of such images demonstrate that intuition should initially run unchecked. Though a general distrust of images in the mathematical community has historically confined such benefits to the arts, Fractal Geometry is a field where visual exploration is essential.

Introducing fractals without touring their images would be like introducing a beautiful symphony by displaying only the score. Like music, the pictures of these sets are free of language barriers; the numbers, formulas, and definitions typical in other areas of mathematics are absent here. One of the best qualities of a fractal is that visually, they are universally accessible. "That fractal reminds me of a..." or "This one looks like one of those..." An unguided tour of fractals allows people to notice these things. It gives us an opportunity to make our own inferences and allows different explorers to see different characteristics. The following pictures are meant as a visual introduction.

Enjoy!

What's in a Word?

But what *are* Fractals? When Benoit Mandelbrot first began to bring together his ideas concerning this new geometry, he was able to make conjectures without a new classification of the objects he investigated. By 1975, however, the shapes that Mandelbrot had encountered and described in his wide scope of mathematical travels required a name. Mandelbrot chose "fractal" and in *The Fractal Geometry of Nature* he states:

> *"I coined fractal from the Latin adjective fractus. The corresponding Latin verb frangere means 'to break': to create irregular fragments. It is therefore sensible— and how appropriate for our needs!— that, in addition to 'fragmented' (as in fraction or refraction) fractus should also mean 'irregular', both meanings preserved in fragment."*

[Mand82]

This name encompassed common characteristics of the objects Mandelbrot had encountered. The Weierstrass functions, the Koch snowflakes, the not-quite filled surfaces of Sierpinski's triangle, and the infinite winding of the coastline were all rough, and regardless of magnification, never attained a smooth uniformity.

Defining Fractals

Once she or he has established a field, the sole job of a mathematician is to further classify the mathematical entries of that field. By naming a new category of objects, Mandelbrot took the first step in classifying the shapes of a new geometry. The next step is to solidify and formalize the definition of the word fractal. However, until a precise definition of the concept of fractals is set, a complete categorization must wait.

In his first book, Mandelbrot avoided giving any formal definition to the word fractal, but in 1977 when *The Fractal Geometry of Nature* was released, Mandelbrot presented the following definition:

> *"A fractal is a set for which the Hausdorff-Besicovitch dimension strictly exceeds the topological dimension."*

[Mand82]

In a revision of the text published in 1982, Mandelbrot voiced his regret at having defined the word fractal. For one, this definition excluded some shapes with fractal-like qualities. Further, Mandelbrot felt that tying the definition of a fractal exclusively to its fractal dimension was premature. In order to understand why he felt this was unwarranted, it is first important to understand the way mathematicians structure geometries.

A useful example is Euclidean Geometry. In Euclidean Geometry there are certain rules which must be followed. Most importantly, only "rigid motion" manipulations of geometrical objects are allowed. Rigid motions are those flips, rotations, and movements of triangles, cones and other shapes, that preserve distance. Movements that will change the distance between points (such as twisting or stretching) are "illegal". Euclidean properties are those characteristics that do not change, or are invariant, under these distance-preserving transformations.

Because Euclidean properties are characteristics unaffected by movements such as flipping and rotating, the fact that the triangle below has one 90° angle both before and after the transformation indicates that this is a Euclidean property of the triangle. In comparison, the fact that the orientation of the triangle is changed by the transformation indicates that orientation fails to meet the definition of a strictly Euclidean property. (See *Figure 15.*)

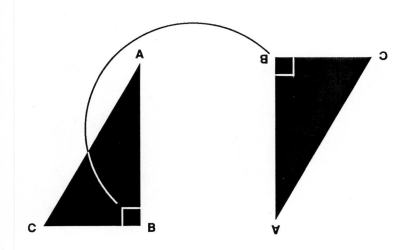

Figure 15 Rigid Motion Triangle

To summarize, in its logical structure Euclidean Geometry is concerned with properties of shapes that are unchanged by those transformations and manipulations that preserve distance. Other geometries are interested in characteristics that are invariant under transformations that preserve angles. Fractal Geometry is interested in all those characteristics that are unchanged under transformations and manipulations that preserve _____. Research must fill in the blank, but Fractal Geometry should ultimately define properties as being invariant under certain transformations.

"I feel — the feeling is not new, as it had already led me to abstain from defining fractals in my first book — that the notion of fractal is more basic than any particular notion of dimension. A more basic reason for not defining fractals resides in the broadly held feeling that the key factor to a set's being fractal is invariance under some class of transforms but no one has yet pinned this invariance satisfactorily."

[Mand83]

Though the name seems to fit the objects well, Mandelbrot and others think it best at this point that the word fractal remain undefined. This does not necessarily hurt the expansion of the field. To the contrary; if adhered to, it can help prevent limitations imposed by a narrow definition. (See *Figure 16.*)

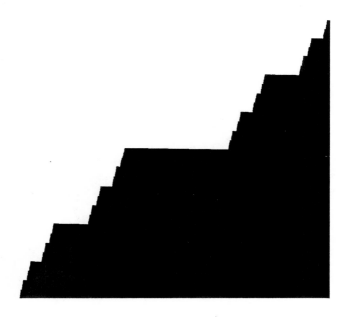

Figure 16 The Devil's Staircase

Though Mandelbrot suggested in 1982 that "fractal" be left undefined, the majority of science writers are unaware that fractals are better off without a strict definition. The most common definition used for fractals in recent literature is "a set for which the Hausdorff-Besicovitch dimension strictly exceeds the topological dimension," (i.e.,1,2,3...). But consider the object in Figure 16, which is the Cantor function, or "Devil's Staircase". Though it certainly fits some fractal criteria, the above definition is too narrow to include it. This is a set for which the Hausdorff-Besicovitch dimension (D_{h-b}) is actually equal to the topological dimension (D_t). Because the sums of the widths and of the heights of the steps both equal 1, and the curve has a well-defined length 2, it is called rectifiable and is of dimension 1. Therefore $D_{h-b} = D_t = 1$. Rather than clarifying the situation, our "definition" actually turns out to be more limiting than it is helpful! [Mand82]

Having no "official" definition for fractals can be frustrating, and excerpts from recent literature show that in the absence of a formal definition, many are still struggling to capture a precise description of these shapes. Though the following quotes came from both popular science and academic periodicals, they help to summarize and highlight the most pronounced features of fractals.

A fractal is... "a set for which the Hausdorff-Besicovitch dimension strictly exceeds the topological dimension."
Peitgen, The Beauty of Fractals 1986

Fractal Geometry is... "a branch of mathematics that can describe and analyze the irregularity of the natural world."
The Economist 1987

Fractal Geometry is... "where the parts resemble the whole, the parts' parts resemble the parts, and self-similarity continues, in some cases down to the atomic level."
La Brecque, Mosaic 1987

Fractals are... "simultaneously well-ordered and wildly chaotic."
La Brecque, Mosaic 1987

Fractals are... "irregular and fragmented shapes which can be magnified endlessly and still retain their complicated structures."
Peterson, Science News 1987

Fractals are... "geometric forms whose irregular details recur at different scales."
Horgan, Scientific American 1988

A fractal is... "a type of object that reveals more detail as it is increasingly magnified."
Dewdney, Scientific American 1986

Fractals are... "figures with fractional dimensions; that is, figures whose effective dimensions exceed their topological dimensions. Fractals differ from ordinary geometrical shapes; they appear rough or fragmented, and this fragmentation exists at all scales. Thus, many fractals could also be called self-similar, although the precise definition of self-similarity is more restrictive; the term means that they are similar— either exactly or statistically— at any magnification."
Jeffery, Byte 1987

"The one essential element of fractals is their peculiar fractional dimensionality... and, in practice, every fractal that has ever been of use has also had self-similarity in one way or another— the small parts look like the big parts— but even that isn't a mathematical requirement."
Sorenson, Byte 1984

Fractals are... "fractional-dimensional spaces... They behave as if they had an intermediate number of dimensions."
Sky & Telescope 1986

Fractal Geometry is... "a branch of mathematics concerned with finding order in apparent randomness."
Science Digest 1986

Fractals are... "shapes that look more or less the same on all, or many, scales of magnification."
The Economist 1987

"The main characteristic is an extreme irregularity that extends with no break and no relative attenuation down to infinitesimal scales."
Mandelbrot, New Scientist 1978

A fractal is... "an object with a sprawling tenuous pattern. As the pattern is magnified it reveals repetitive levels of detail, so that similar structures exist on all scales."
Sander, Physics Today 1985

*Fractals are... "temporal or spatial phenomena that are
continuous but not differentiable and that exhibit partial
correlation over many scales. They refer to a series in which
the Hausdorff dimension exceeds the topological dimension.
A continuous series is differentiable because it can be split
up into an infinite number of absolutely smooth straight
lines. A non-differentiable continuous series cannot be
resolved. Every attempt results in still more roughness."*
Burrough, Nature 1981

*"What are fractals? Different people use the word fractal in
different ways, but all agree that fractal objects contain
structures nested within one another like Chinese boxes or
Russian dolls."*
Kadanoff, Physics Today 1986

*Fractals are... "a new language for the description of nature,
supplanting the simple geometric figures that have been used
up to now."*
Thomsen, Science News 1987

*Fractals are... "curves and surfaces that live in an unusual
realm between the first and second, or between the second
and third dimensions."*
Thomsen, Science News 1982

*A fractal is... "one whose detailed structure is a reduced
scale (and perhaps deformed) image of its overall shape."*
Mandelbrot, Proceedings Int'l Congress of Mathematicians 1983

*"Examples of fractal structures have the property of scale
invariance: that is, they 'look' the same (in a statistical
sense) in a photograph of any resolution. An important
property that characterizes them is an effective
dimensionality, the Fractal Hausdorff Dimension D, which
need not be an integer."*
Sander, Physics News 1984

"Fractals have a property called self-similarity— that is they have similar features at all length scales and therefore look the same at all magnifications— and are characterized by effective fractional dimensions, rather than at the 1, 2, and 3 of curves, surfaces, and volumes."
Robinson, Science 1985

"The concept of fractals deals with complex geometric forms. A fractal structure is not smooth and homogeneous when examined with stronger and stronger magnifying lenses; fractals reveal greater and greater levels of detail. The smaller-scale structure is similar to the larger-scale form. There is no characteristic scale of length."
West, Goldberger, American Scientist 1987

Fractals are... "extremely irregular curves or surfaces that live in a realm somewhere between the first and second dimension, or between the second and third dimension."
Thomsen, Science News 1982

Fractals are... "those lacy snowflaky figures with dimensions somewhere between one and two, or two and three dimensions."
Thomsen, Science News 1980

Fractals... "mathematically justify tie dye."
Marc Ratner, 1990

"Distinguishing fractals from the geometric figures Euclid taught us about are two basic qualities: self-symmetry and fractional dimension."
Thomsen, Science News 1987

THE SPIRIT OF FRACTALS

The theory of Fractal Geometry is a mere 15 years old and a precise definition of the term "fractal" remains elusive. An interesting alternative was offered by Michael Barnsley of Georgia Tech when he referred to the "spirit of fractals." As can be seen in the objects that inspired Mandelbrot, "in spirit," fractals can have at least two distinguishing characteristics: fractal dimension and self-similarity. (See *Figures 17 & 18.*)

Figure 17 Fractal Dimension

The figure on the left has a fractal dimension close to 2. However, the more jagged figure on the right has a fractal dimension closer to 3.

Figure 18 Self-Similarity

Fractal Dimension

> *"Fractal dimensions can be attached to clouds, trees, coastlines, feathers, networks of neurons in the body, dust motes in the air at an instant in time, the clothes you are wearing, the distribution of frequencies of light reflected by a flower, the colors emitted by the sun, and the wrinkled surface of the sea during a storm. These numbers allow us to compare sets in the real world with the laboratory fractals."*

[Barn89]

One of the most important qualities of fractals discovered thus far is fractal dimension. Without an intuitive understanding of the Hausdorff-Besicovitch dimension, "Fractal Geometry," Mandelbrot later remarked, "would not be" [Peit86].

Though fractal images themselves seem intuitively familiar, the concept of fractal dimension is not. How then did Mandelbrot, and more importantly how can we intuitively grasp the apparently non-intuitive idea of fractional dimensions?

It is important first to understand that in the evolution of this new type of dimension, these "in between" values did not replace, but rather augmented the tools available for classifying objects. The dimension crisis at the turn of the century had forced an expansion in the theory of dimension to include numbers between 1, 2, 3, etc. Hausdorff established the theoretical basis allowing for such values. Mandelbrot, however, was able to identify both mathematical and physical examples that possessed such dimensions. Since that time, mathematicians, physicists, and scientists have come up with a variety of numbers, some theoretical and some experimental, that are associated with both natural fractals, and with those from the mathematical laboratory.

Between Dimensions

> *"There are various numbers associated with fractals which can be used to compare them. They are generally referred to as fractal dimensions. They are attempts to quantify a subjective feeling which we have about how densely the fractal occupies the (metric) space in which it lies."*

[Barn89]

Our conventional definition known as topological dimension continues to be valid and classifies Euclidean shapes as points, lines, surfaces, and solids, with respect to length, area and volume. By definition, their dimension must be a whole number, and topologically objects have dimension 1, 2 or 3. (See *Figure 19.*) The theory worked out by mathematicians allows us to expand this idea, and, in addition to an object's topological dimension, we can study its fractal dimension as well.

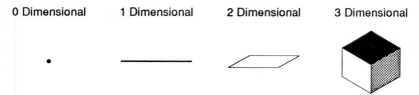

0 Dimensional   1 Dimensional   2 Dimensional   3 Dimensional

Figure 19 Euclidean Dimension Examples

Like topological dimension, the fractal dimension helps describe and categorize objects. The fractal dimension "need not be an integer, and quantifies an object's ability to fill the topological space in which it is embedded " [LaBr87]. This puts fractals in a realm "between dimensions"; whereas Euclidean shapes fall into three categories: a line, a surface, or a solid; fractals may lie somewhere in between. The fractal dimension measures the roughness, or space filling ability of an object. As one writer put it, "Thus in a 3-dimensional setting a fractal with dimension 2.9 would be somewhat like a sponge, while a fractal with dimension 2.3, less space filling, would be merely rough hewn" [LaBr87].

If we look back at the Koch example we can see that after an infinite number of generations, the boundary of the snowflake will be infinitely jagged and bumpy. The boundary is so convoluted that it takes up "more space" than a line, and yet it still does not solidly fill a plane. Its fractal dimension is somewhere in between one and two and, as stated earlier, is estimated to be about 1.2618.

Another example is the Menger sponge. It starts out as a solid of dimension three, then by a recursive algorithm the middle ninth of each cube is removed. As this continues, ad infinitum, we can think of the Menger sponge as "losing" more and more of its volume; until it lies somewhere between a surface and a solid. In the strictest sense, its 2-dimensional "volume" is infinite, and its 3-dimensional "volume" is 0. For us, this means that the Menger sponge ends up having infinite surface area and zero volume. If you put the Menger sponge in a glass of water, the water level would not rise at all (as it does when you put a rock in it), which illustrates the idea of 0 volume. However, you could never paint all of the Menger sponge because it has infinite surface area.

To answer a very common question, outside the mathematical realm we cannot execute these algorithms an infinite number of times. Our eyes only allow us to see so much detail, and these pictures are shown at a level of iteration just sufficient to give their fractal flavor. The range over which these sets are aesthetically pleasing corresponds in some ways to the range over which natural objects are also recognizably fractal [Hort88].

There are abundant fractals, both natural and mathematical that exhibit these *in between* dimensions. (See *Figure 20*.) In the above examples, the Koch snowflake was between one and two and the Menger sponge was between two and three. Parallel to the process that creates the Menger sponge, we can see that if we start with a line and successfully remove more and more of it, the resulting Cantor set will be our example for an object between zero and one. Calculation confirms this, and the fractal dimension turns out to be approximately d=0.6309.

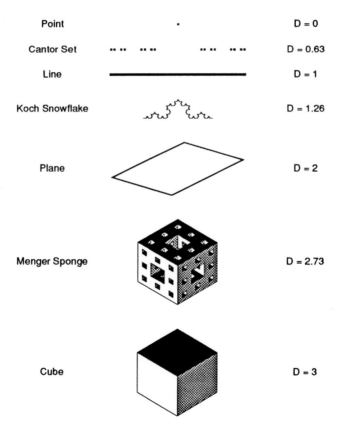

Point	D = 0
Cantor Set	D = 0.63
Line	D = 1
Koch Snowflake	D = 1.26
Plane	D = 2
Menger Sponge	D = 2.73
Cube	D = 3

Figure 20 Fractal Totem Pole

> *"A crinkly line of, say, 1.25 dimensions is better at filling up space than a one-dimensional straight line because you need more ink to draw the crinkly than you do to draw the straight line. A line of 1.26 dimensions is even crinklier and needs even more ink."*

[Econ87]

Since people are likely to have more experience describing lightning than a Koch curve, natural fractals may be more helpful than mathematical constructs for explaining fractal dimension. Take for example lightning: if we compare a picture of a streak of lightning to a picture of a straight line, the lightning seems to take up more space. If the lightning was a straight line, its dimension would be 1. But it is not. On the other hand, if it solidly filled the picture, its dimension would be 2. But it does not. Its zig-zagged pattern fills up more space than a line would, yet it doesn't fill the entire plane. In fact, in this sense its effective fractal dimension is estimated to be about 1.4 [Garc88].

Another example is the surface of a mountain. It is important to distinguish between the surface of the mountain, and the mountain itself. While the entire mountains themselves are three-dimensional, their surfaces only are considered two-dimensional. If the surface of a mountain were as flat as a pancake it would have dimension 2. In the fractal sense, however, the rugged surfaces of mountains are crumpled enough to lie somewhere in between the second and third dimensions. The surface of a bumpy hill might have dimension 2.1, while that of a jagged mountain range would be higher, somewhere around 2.6.

The fractal dimension of an object is a measure of space filling ability and allows us to compare and categorize fractals. Objects, that until this point have necessarily been described as crinkled and rugged, now have an associated numerical value that helps quantify these qualities. (See *Figure 21*.)

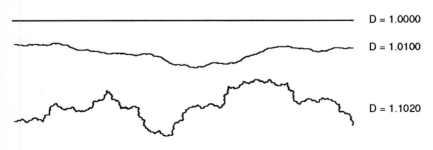

D = 1.0000

D = 1.0100

D = 1.1020

Figure 21 Crinkly Lines

Figure 22 Magnification: On the Boundary of the Mandelbrot Set

Discovered by Mandelbrot in 1980, this set is one of the most complex objects in mathematics today. Infinite worlds of beautiful, swirling, wildly fractal shapes lie in the magnifications of its boundary. Because it is so complicated, this boundary is expected by some to have a dimension actually equal to 2. But after ten years this conjecture has remained unproven [Hort88].

Calculating Dimensions

Popular as well as scientific literature readily quotes fractal dimensions when discussing fractals. But how do mathematicians and scientists arrive at these numbers? The answer to the question can be difficult, and as one popular magazine said "the best way to understand fractal dimension is not to worry about it" [Econ87]. The point of the writer is understood, and the beauty of fractals does lie in the fact that "one does not need to understand a formula in order to understand the concept of fractals" [deMi90]. However, the best way to understand math is not to ignore or avoid it, as many of us are so often told to do. The best way is to jump in, hopefully with guidance, and rather than allowing vocabulary of the mathematical language to block the path to learning, learn to skim past it in order to focus on the flow of logic behind a concept.

So, rather than put this section in a mathematical appendix, tucked somewhere in the back of the book, easy to avoid, I placed it here. As a product of our mathematical education system, I want to point out that, unfortunately, along with my degree in math came the ability to feel very comfortable about not understanding a concept. Though there are more preferable ways of learning, prolonged bewilderment by a subject does not have to be a permanent roadblock. Understanding does not always come instantly, it takes time.

Initially, most of us feel lost the first time we have to read another language. Mathematics is no different. With a particularly dense subject, vocabulary is important, as is an overview of the topics to be covered. The following two sections discuss two methods of actually calculating fractal dimension. The first part will talk about a method called "box-counting", and the second generalizes a technique for calculating fractal dimension used by Mandelbrot in *The Fractal Geometry of Nature*. These sections use vocabulary words such as the "logarithmic function" (abbreviated as ln(x), read as "the natural log of x"), the "slope" of a line, and "limits." (See *Figures 23, 24 & 25.*)

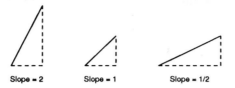

Slope = 2   Slope = 1   Slope = 1/2

Figure 23 The Slope of a Line

The slope of a line is a measure of its steepness. For example, the line representing the side view of the very steep hill will have a higher slope than that of the gentler hill.

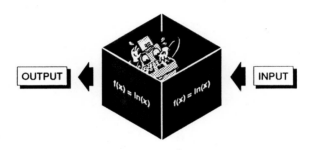

Figure 24 Box Representing Function

In the simplest sense, the natural log function, abbreviated as ln(x), can be thought of as a box similar to that above; values are put in to the box, acted upon according to a rule, and then sent out the other end.

A   B

Figure 25 Limits

Mathematicians love to know what will happen after an infinite number of mathematical operations have been executed. However, since nobody has time to wait around, they use reasoning to deduce the net result after an infinite number of mathematical manipulations. This net result is called the limit, and in calculus, "taking the limit" of a function means to find out, based upon specific circumstances, what the function will eventually do. The man above is walking half way from point A to point B. Then he walks half way from his new position to point B. Will he ever reach point B? Not in our life time. In fact, he won't reach point B until an infinite amount of time goes by. We can however *deduce* that he will ultimately reach point B and we say that point B is the limit.

Box-counting and Log vs. Log Graphs

There are several ways of determining a fractal dimension, including experimentation in the case of "natural fractals" and derivation in the case of mathematical fractals. Using experimentation, many scientists have determined fractal dimensions by recognizing a certain type of relationship between two characteristics of a system. This relationship is plotted on a special graph. Normally, mathematicians plot the values of x vs. y on a graph, but in this case they plot the logarithm of x, vs the logarithm of y. This is called a log vs. log graph. The slope (steepness) of the resulting line is the fractal dimension of the system in question. In what is often called the box-counting method, this idea is used to determine the fractal dimension of "real-world" entities, including "rivers, clouds, coastlines, trees, or villi of intestinal walls" [Jurg90]. (See *Figure 26*.)

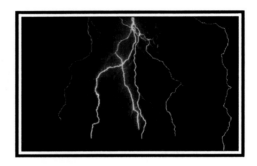

Figure 26 The Box Counting Method

Photo Courtesy of Claire McDowell. Bald Mountain Lookout, Sequoia National Forest, September 1984.

The box counting method assumes that if one were to take, for example, our picture of lightning, the number of boxes necessary to "cover" the shape would increase as the length of the sides of the boxes decreases. (Simple enough; the smaller the boxes, the more we need to cover the shape.) We can use an equation to precisely describe this relationship by saying that the number of boxes, $N(l)$, of sidelength 'l,' will vary as $N(l) = (l/l_o)^{-d}$ (where l_o is unit length) [Frae86]. After some algebraic manipulation, d is restated as a relationship between $\ln(N(l))$ and $\ln(l/l_o)$. If we plot these data points on log-log paper, the slope of the resulting straight line approximation is the fractal dimension of the object surveyed.

The slope of such a graph is constant for exactly self-similar fractals. That is, the "steepness" doesn't change and the method, therefore, yields one value for the fractal dimension of the shape. However, in some cases, especially for models of natural phenomena, these fractals may not be exactly self-similar. They are often what we call self-affine fractals, and for such fractals the slope of their log-log plot changes. "More often than not, for natural fractals, this slope can vary over the length scales sampled, in which case the resulting accrete is referred to as a self-affine fractal" [Lakh87]. These fractals may be characterized by many fractal dimensions [Mand86].

This brings to light another area of recent study; that of multifractal phenomena. "Multifractal phenomena" according to H.E. Stanley of Boston University, "describes the concept that different regions of an object have different fractal properties" [Meak88]. Behavior such as fluid flow in porous media (such as oil in porous rock), and that of complex surfaces and interfaces (such as the layers in the atmosphere), "seem to require not one, but many exponents [dimensions] for their description" [Meak88].

The Size of Fractals

Early in the dimension crisis, when mathematicians investigated the size of fractals, they often derived results of zero or infinity for common parameters such as length or area. In the case of purely mathematical constructs, a very common approach for deriving the more meaningful fractal dimension is one taken by Mandelbrot in *The Fractal Geometry of Nature*. The following is a generalization of that method. The emphasis is placed on finding the d-dimensional space in which the "size" of the fractal is neither zero nor infinite when approached with standard methods of measurement.

Let S_d = "size" of object in d-dimensional space. Then, for example

S_1 = "size" of object in 1-dimensional space = Length

S_2 = "size" of object in 2-dimensional space = Area

S_3 = "size" of object in 3-dimensional space = Volume

Let n = generation number

N = number of pieces in the nth generation

l = length of each piece in the nth generation

To find out the "size" of a shape we usually begin by taking the number of parts the shape is cut into, and multiply this by the size of the parts. For example the size, or perimeter of your square living room would be the number of walls, 4, multiplied by the length of one wall, say 10 feet. Therefore, the size, or perimeter of your living room is 40 feet. It makes sense to say that the "size" of a fractal can be calculated the same way: counting the number of segments in the fractal and multiplying that by the size of each segment.

Therefore, it is logical to define the total size as the product of the number of pieces of the fractal and the size of these pieces:

$$N \cdot l = a$$

As Richardson and Mandelbrot pointed out, the "size" of a fractal depends on two things: its fractal dimension (d), and the number of generations, or iterations involved in its construction (n). Therefore, we define the size as

$$N \cdot l^d = a_{(n,d)}$$

where the number a depends on both the changing n and the power d for its value. But, since fractals are the result of an infinite number of generations, let "n," the number iterations, do just that; let n "go to infinity" as mathematicians say. Then we may define the size S_d as the limit, or "net result" after an infinite number of generations. Translated into "math" this is written in the following way:

$$S_d = \lim_{n \to \infty} a_{(n,d)}$$

We can define d as the number for which we don't get an answer of zero or infinity for the size of the fractal:

$$0 < \lim_{n \to \infty} a_{(n,d)} < \infty .$$

Or, by direct substitution, *d is the exact number for which the "size" of the object in its d-dimensional space makes sense (not = 0, ∞).*

$$0 < S_d < \infty$$

It turns out that there is such a number d, and that choosing numbers less than d gives infinite size, and choosing numbers greater then d gives zero size.

The Koch Snowflake

A classic example used for computing fractal dimension is the Koch snowflake. The size of the snowflake is defined as the product of the number of pieces of the snowflake multiplied by the size of each piece. After only two iterations, a general pattern develops for determining this size.

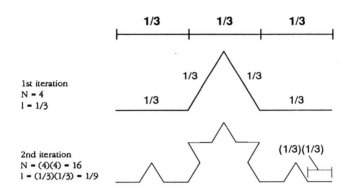

Figure 27 Calculating Dimension

After the first generation, n = 1, and the total length is $4^1/3^1$. When n=2, the length is $4^2/3^2$. Therefore after n generations $N = 4^n$ and $l = (1/3)^n$ and the length, $(4/3)^n$, as in the coastline example, is infinite (as n gets closer to ∞, $(4/3)^n$ also goes to infinity). To pursue the dimension, substitute into the general formula

$$N \cdot l^d = a_{(n,d)}$$

$$4^n \cdot (1/3^n)^d = a_{(n,d)}$$

Then, in order to derive the dimension, find the unique d for which the size is in between zero and infinity.

$$0 < s_d < \infty.$$

$$0 < \lim_{n \to \infty} a_{(n,d)} < \infty.$$

Solving

$$4^n \cdot (1/3^n)^d = a_{(n,d)}$$

Using logarithmic functions and their properties

$$\log (4^n \cdot (1/3^n)^d) = \log (a_{(n,d)})$$

$$\log (4^n) + \log ((1/3^n)^d) = \log (a_{(n,d)})$$

$$n\log 4 + d\log (1/3^n) = \log (a_{(n,d)})$$

$$n\log 4 + d\log 1 - d\log (3^n) = \log (a_{(n,d)})$$

$$n\log 4 + d\log 1 - dn\log 3 = \log (a_{(n,d)})$$

dividing by n

$$\log 4 + d(0) - d\log 3 = (1/n)\log (a_{(n,d)})$$

Now, taking the limit

$$\lim_{n \to \infty}[\log 4 - d\log 3] = \lim_{n \to \infty}[(1/n)\log(a_{n,d})]$$

Since lim (1/n) = 0, then

$$\log 4 - d\log 3 = 0$$

$$\log 4 = d\log 3$$

$$\log 4/\log 3 = d$$

$$1.2618 = d$$

Again d is the only number for which the "size" in d-dimensional space makes sense. It is easily shown (I hate when they say that!) that a value of d = 1 in the Koch example will yield a "size" of infinity for this 1-dimensional volume, which we call length, and a value of d = 2 will yield a "size" of 0 for what we will consider area. Only one d yields a size between zero and infinity. This is parallel to the concept stated earlier for the Menger sponge: it has infinite surface area and zero volume.

Thus, we have a firm theoretical basis when we say that a fractal curve has infinite length, yet zero area, and lies somewhere between a line and a plane. In the case of the Koch snowflake, its size makes sense only in its 1.2618-dimensional space. Now we can understand why, in one of his first articles on Fractal Geometry, Mandelbrot says, "Thus all coastlines are infinite, but coastlines with high values of d are more infinite than coastlines of smaller d" [Mand78].

"The" Fractal Dimension

As with any field where the theory has yet to be solidified, avoiding common pitfalls is both tricky and necessary. Probably the most frequent error is a result of confining ourselves to definitions as quickly as possible, regardless of the accuracy of the definition. This has been a problem in both theoretical and experimental research concerning fractal dimension.

In one area, mathematicians continue to address and expand new types of dimension, thereby broadening the scope of that concept. In the Koch example we demonstrated just one way to calculate one aspect of fractal dimension. It is important to stress that different methods may yield different dimensions, just as the study of different aspects of dimensionality (topological or fractal) may yield different values. In fact, recent studies focus on the case of multifractals which may have an infinite number of fractal dimensions.

Thus, to avoid the common error of labeling one value as "the fractal dimension" to the exclusion of another acceptable value, we should heed Mandelbrot's call for a more inclusive, generic definition of fractal dimension. In a last minute addition to a printing of *The Fractal Geometry of Nature*, Mandelbrot warns those involved in either effort.

> *"To my chagrin, the term 'Hausdorff dimension' has started being applied indiscriminately to either of the dimensions Hausdorff-Besicovitch or Frostman, and to further variants thereof. Other writers go to the opposite excess: they overstress the methods most often used to estimate D in practical work, such as the similarity dimension, the exponent in the mass-radius relation, or a spectral (experimental) exponent, and they proceed to enshrine them to define 'the' fractal dimension."*

> *"It is a pity that most of these reactions manifested themselves a bit too late. They would have encouraged me to return in the present book to the well-inspired approach taken in 1975: to leave the term 'fractal' without a pedantic definition, to use "fractal dimension" as a generic term applicable to all the variants, and to use in each specific case which ever definition is the most appropriate."*

[Mand82]

The establishment of Fractal Geometry has expanded the mathematician's tool-box for description, and has sparked interesting new research, both applied and pure, on dimensionality. In their quest to define these dimensionally ambiguous objects, scientists continue to formulate generalizations about this fractal dimension and the role it plays in fractal theory.

Sometimes, regardless of how much reading, listening, or watching we try, we actually need to do, in order to understand. So...look at the next page. Run the palm of your hand over the paper. The part that is touching your palm is the surface of the paper. This is two-dimensional, and for the purpose of this exercise we assume it has no thickness.

Now, rip the page out and crumple it into a tight ball (Really!). Unwad the paper and run your palm over the surface of the page again. The surface is bumpier, rougher, and somehow seems to take up more space than it did a minute ago. The surface (the portion that tries to touch your palm) seems to be more than two-dimensional. It does not have volume, however, and therefore is still not three-dimensional. In the fractal sense, the surface of the paper lies somewhere in between the two dimensions.

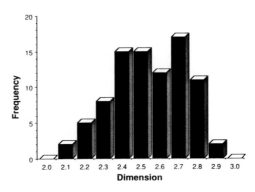

Crumpled Paper Balls

The calculation of the fractal dimension for this example is given to first year physics students at UFPE in Brazil. The surface of the paper "fills up more space than a plane" yet is still not a solid. The average dimension is about 2.5 and is determined experimentally from $L=km^{1/d}$. Here L is the average diameter of the paper ball, m is its mass and $k (1/p)^{1/d}$, where p is the average mass-density on the fractal structure [Gome86].

Rip me out!

> *"For me, the most important instrument of thought is the eye.*
> *It sees similarities before a formula has been created to identify*
> *them."*

B. Mandelbrot

As with many facets of Fractal Geometry, the best way to introduce the concept of self-similarity is through the sense of sight. It is said that we can process up to 10 times more information visually than we can through any other medium. Now with the aid of the computer, the field of computer graphics, and the incredible world of fractals, we can begin to utilize the underdeveloped resource of vision in the mathematically trained mind.

H.O Peitgen states,

> *"We can discover patterns and connections. We can make*
> *conjectures by using our visual capabilities, which is something*
> *that has not been developed in math brains so far."*

[LaBr87]

Self-similarity, like fractal dimension, is a distinct characteristic of the science of Fractal Geometry. The ability to recognize fractals lies partly in fine-tuning our ability to recognize this self-similarity in the world around us. With this in mind, notice that the following fractals exhibit this characteristic through a recursive, "Russian-doll" style of "whorls within whorls" and "ferns within ferns."

Repetitive Structure

Because self-similarity is now so visually apparent, it may seem to have taken the mathematical community an unwarranted amount of time to identify self-similarity. But by the late 17th century Leibniz had already recognized this property in the line, and physicists have for a long while known of the scaling patterns in the classic study of Brownian motion, (which is "the movement of tiny particles in a liquid caused by the motion of the liquid's molecules") [Batt85]. What has taken so long, as Mandelbrot pointed out, is that though scientists have addressed such phenomena analytically, they missed the corresponding geometric properties, and thereby overlooked their undisputable link with nature [Mand78].

Though many objects in nature may appear unstructured, they in fact exhibit a repetitive pattern on finer scales. In contrast, this repetitive pattern, or scaling, is often absent in rather simplistic objects. The familiar shapes of Euclidean geometry, for example, often lose their structure

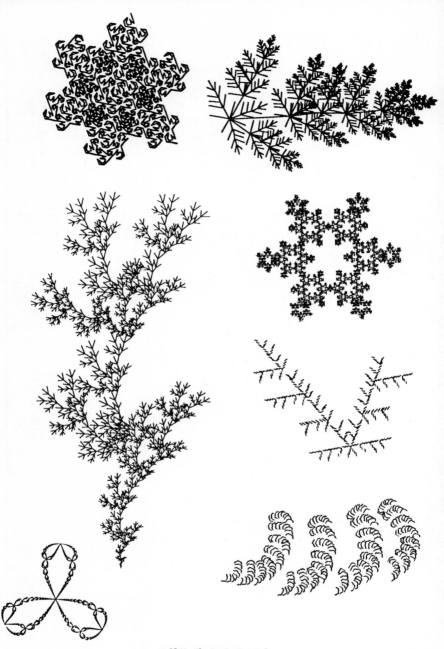

Self-Similarity in Fractals

"Fractals tell us to look at an object at all levels, because there are interesting things going on everywhere."

Oppenheimer [LaBr87]

when magnified. "The surface of a large sphere appears almost flat when viewed close up, which is why plenty of people used to think the earth was flat" [Econ87]. A fractal, however, is just the opposite; it maintains its geometric structure over many, and perhaps an infinite number of scales. The study of such objects has the advantage that, unlike the example of the earth's surface, properties discovered on finer scales may be characteristic of the object as a whole, and vice-versa. Scientists often refer to this phenomena as dilation symmetry, which means that fractals "look geometrically self-similar under transformations of scales such as changing the magnification of a microscope" [Scha89].

When an object maintains its structure over a range of scales, this is called scale invariance. Recent studies show that "a variety of natural processes exhibit scale invariance over a wide range of scales" [Turc86]. This characteristic is a flag signaling the presence of fractals that has been found across the range of the physical, biological, and social sciences as well. Therefore, in addition to giving information about the structures themselves, further research into the presence of scaling may shed light on the similarities of the processes that result in these fractal shapes. For this reason, it is useful to investigate different degrees of self-similarity and to search for a way to begin categorizing fractals based on these differences.

Categorizing Self-Similarities

A distinction can be made between two kinds of self-similarity: exact self-similarity (linear fractals), and statistical self-similarity (non-linear fractals). The simplest cases are those that are exactly self-similar. Due to their repetitive construction, both the Sierpinski triangle and the Cantor set are examples of such fractals. The same manipulations are applied to, in this case, successively smaller portions of the objects, and therefore this exact self-similarity is built in. In the design of the Cantor set "remove the middle third of a line segment" is the instruction that is repeatedly applied; for the Sierpinski triangle the instruction is "remove an inverted triangle." Altering these algorithms results in interesting variations. However, as long as these instructions are consistently reapplied, the new objects maintain this exact self-similarity. (See *Figure 28.*)

Figure 28 Mutant Sierpinski Triangle

The figure on the left is the "normal" Sierpinski Triangle. The figure on the right uses a slightly different algorithm.

Alternatively, fractals with such a high degree of self-similarity can be thought of as being composed of shrunken images of themselves. The Sierpinski triangle, for example, contains an infinite number of smaller Sierpinski triangles throughout its structure. For natural objects that are exactly self-similar over a sufficiently large range, this approach can make computer generation of their images easy and surprisingly life-like. When the image can be made up of smaller replicas of the entire shape, a relatively small amount of information can describe a rather complicated natural object. (See *Figure 29*.) As Glenn Zorpette points out in the IEEE Spectrum:

> *"Because computer-generated fractal images have similar patterns on many different scales, relatively little code is all that is usually needed to create them. Once written to produce the detail on one scale, much the same software can be reused in a loop to repeat the image on successively larger (or smaller) scales. Thus a remarkably intricate image blossoms from a small, simple piece of software."*

[Zorp88]

Figure 29 Black Spleenwort Fern

> *"The very regular things are the easiest fractals, but there are much more complex things that are essentially fractal, and which are more interesting in the long run."*

A.R. Smith, Pixar

Though exact self-similarity is a distinguishing factor in fractals, it is not a prerequisite. As Mandelbrot says, "The word similar does not always have the pedantic sense, 'linearly expanded or reduced,' but it always conforms to the convenient loose sense of 'alike'" [Peit86]. There are those fractals, like the Mandelbrot set or crumpled surfaces resembling mountains, that do not look exactly alike on each scale. However, they do look somewhat alike. Most often these are fractals that we call statistically self-similar.

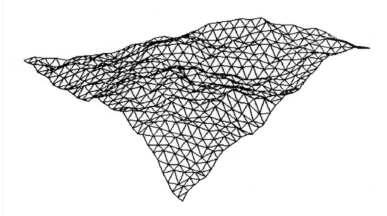

Figure 30 Crumpled Surfaces

As an example of this statistical self-similarity, consider the crumpled surfaces of computer generated mountains. (See *Figure 30.*) It turns out that when adjusted by the appropriate scale factor, the standard deviation (the range around the average height) of any square patch of the crumpled surface is the same as any other patch of that same surface. Though not exactly self-similar, it is statistically self-similar.

The now familiar coastline example also demonstrates statistical self-similarity. A coastline has a variety of bays, and peninsulas. Examining the coastline more closely however, reveals that these bays and peninsulas contain further inlets and protrusions. Though these smaller features are not exact replicas of the larger ones, they have similar, scaled down statistical properties.

There are further variations on the idea of straightforward self-similarity. The fern in *Figure 29* is a classic example of a self-similar fractal. But what of the fern below? Exact self-similarity breaks down at an early point, yet appears again at finer scales. The fern has some self-similar properties, but is not an exact replica at every scale. (See *Figure 31*.)

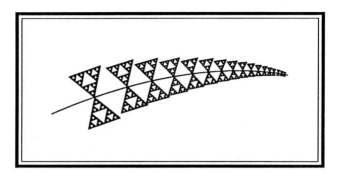

Figure 31 Sierpinski Fern

Because it is comprised of reduced copies of the entire structure, constructing the classic artificial fern is straightforward. However, generating the other "ferns" is slightly more difficult. The study of such objects was the prime focus in a recent class given by H.O. Peitgen at the University of California, Santa Cruz. The attempt to classify fractals based on these varying degrees of self-similarity is a field deserving of research. As Peitgen pointed out to his students, a relatively small amount of research has been devoted to such classifications. "The field" he said "is wide open."

Self-Similarity, Scaling and Physical Phenomena

> *"Self-similarity lets fractal geometers see a sort of order in apparent chaos."*

[Econ87]

Much energy is spent trying to obtain information about those self-similar systems that already exist in the natural world. But what of using human made objects to imitate this characteristic of nature? Mandelbrot points out that from a constructive or engineering standpoint, much could be said for intentionally arranging material in a self-similar manner. If we were to arrange an object in this way, what would the consequences be? Ray Orbach, a physicist at UCLA, Shlomo Alexander of Hebrew University, and Ora Entin-Wohlman of Tel Aviv University recently considered these same questions.

Their work centered on theories concerning thermodynamics and transport properties that could ultimately provide "stronger and safer wings on jet airplanes, as well as more mundane benefits such as cooler handles on cooking pots" [Kozl86].

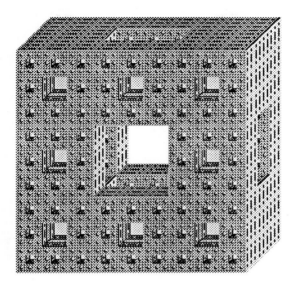

Figure 32 Menger Sponge (D = 2.7268)

Though outside the mathematical realm it is impossible to fully carry out the algorithms necessary to build the above fractal, Mandelbrot nonetheless advocates the study of even the pre-fractal versions of such structures. After describing the sponge, he says, "In engineering, [using] iterated pre-fractal structures may be impractical because of cost, but it seems they would be extremely advantageous, and their dynamics deserve investigation" [Mand78].

As an oak develops from a tiny acorn, a fractal can be generated from a simple seed. The seed is the basic shape of the fractal. We grow the fractal by starting from the basic shape and creating more complex objects in successive generations. The true fractal is reached after an infinite number of generations and is infinitely complex. Early generations often give good approximate pictures of the fractal. [1]

The first fractal in the garden is a group of cacti. It starts from the following seed shape.

Each line segment is replaced by the whole seed in successive generations. The arrowhead denotes orientation. We create the following four generations.

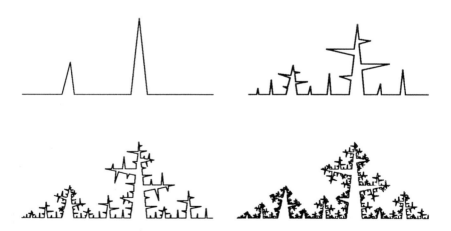

---

[1] This garden is created with the program FractaSketch™

Our garden is grown with straight lines. This is not as strong a limitation as you might think — it is easy to grow curved fractals starting only from straight lines. This fractal and some of the next ones demonstrate that. The seed of this fractal is a triangle. [2]

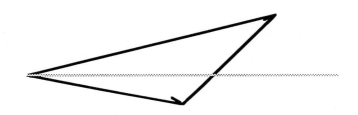

It goes through the following stages of growth.

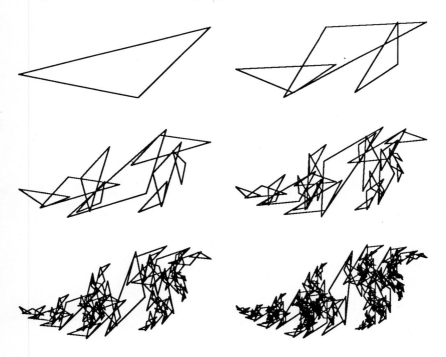

Note how the lines arrange themselves in a continually more curvy manner. We finally get the following wavy shape, which is shown on the next page.

---

[2] The light gray line lets us draw figures that are disconnected from each other.

Fractal Schooner

A seed can be of many different sizes, but it is not necessary to use a complicated seed to get fractals of astounding complexity. Very simple seeds often suffice. For example, take the following simple seed.

We grow a boat with it.

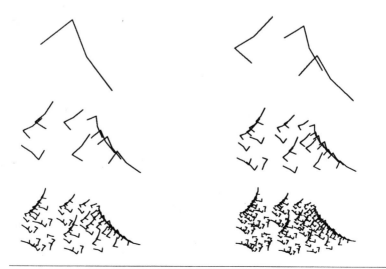

This is the resulting boat with its fractal sails.

A Spinning Dragon

The dragon that lives in our garden has humble origins, and yet flies off to regions unknown. This is the dragon's seed.

The shape that grows from this seed looks completely different.

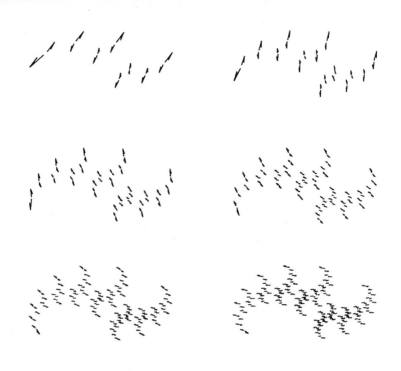

This is the full-grown dragon.

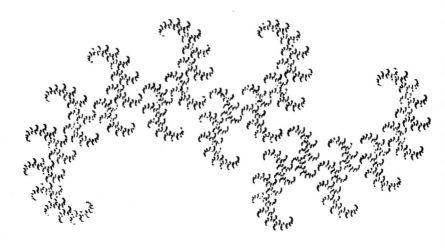

The Fern

A common picture in books on fractals is the fern. We can grow one in our garden too. Start with the following seed.

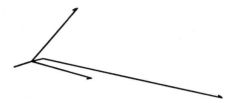

We get these growths that look more and more like ears of wheat. The resulting fractal, however, is a fern.

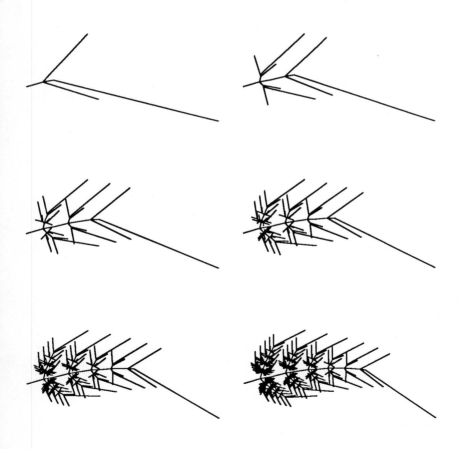

The full-grown fern is a marvel to behold. [3]

Tree

Next to a fern, a tree is one of the most fractal of plants. To obtain the greatest exposure from air and sunlight, trees grow a fractal structure as they strive to fill space with their branches. Our garden is two-dimensional, so we will show a two-dimensional tree. Start with the following seed.

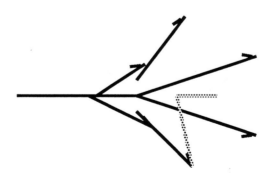

---

[3] The fern is generated until a 5-pixel cutoff is reached with line thickness proportional to length.

This seed grows into a tree.

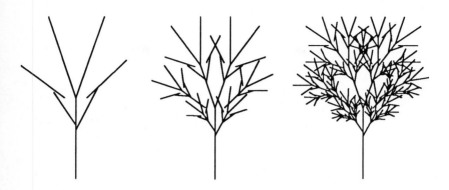

The final tree looks like this. [4]

---

[4] The tree is generated until a 6-pixel cutoff is reached with line thickness proportional to length.

A Spiny Embryo

This seed is a simple geometric figure that results in a complex fractal. Therefore the seed's every detail has a strong effect on the resulting picture. If the seed is symmetric, the fractal will be symmetric, too. A slight asymmetry in the seed becomes amplified manyfold in the final fractal. For example, take this seed.

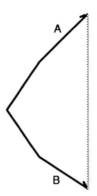

The top arrow (A) is slightly longer than the lower one (B). What effect does this have on the final shape? Let's see.

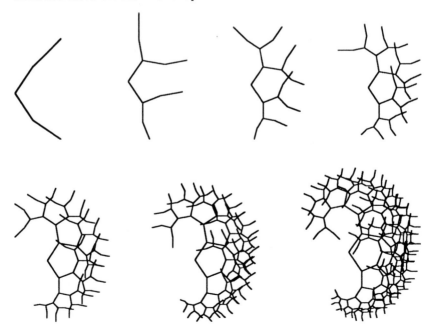

This is the final drawing. The slightly longer top arrow in the seed results in the top of the fractal being much longer than the bottom.

Intricacies

This fractal is an example of an intricate object that has a compelling mixture of order and randomness. It is hard to create fractals with just the right proportions of order and randomness to be pleasing to the eye. In that sense, creating fractals from seeds is like composing a musical score: doing it to its best effect requires mastery. Lots of practice and examples of good pictures are needed. For example, consider this seed. [5]

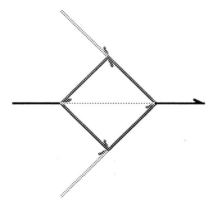

---

[5]The dark gray line inverts the picture when drawing. The light gray line lets us draw disconnected figures.

It grows into an interesting shape.

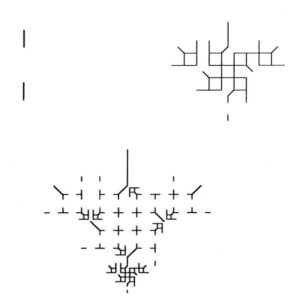

This is the full-grown object.

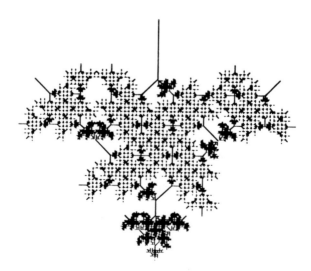

The fractals in this garden show demonstrate the variety of shapes that can result by using the simple recipe of replacing line segments with miniature copies of the whole shape. We have only just begun to explore the richness of the world of Fractal Geometry.

FRACTAL APPLICATIONS

INSTANT COFFEE AND ARTHROPOD BODY LENGTHS

Recently an article appeared in the Journal of Food Science that characterized instant coffee particles using Fractal Geometry. In another area, a scientist compared the fractal dimension of various plants to the body lengths of insects found on the plants. Though these are somewhat specialized examples, people are using fractals in a variety of ways to learn more about the world around them. But why Fractal Geometry? Though the theory is a mere 15 years old, a survey shows fractal-related research to be ever increasing. Fractals are not a catch-all for mathematical modeling, but they are a catch-a-whole-lot.

Why fractals? Fractals seem to exist on that line between order and chaos, as does nature itself. Very often, what had seemed beyond any patterned description is actually ordered, and ordered according to the language of this new geometry. The goal of applied mathematics is to find accurate models of natural processes. Because of this delicate balance between order and chaos, fractals turn out to be excellent mathematical models.

Better Than Average

We make decisions based on mathematical averages almost every day. The average rainfall of an area might be important when choosing a site to grow a certain crop. Or, if the average prices at a certain store are higher than those at another store, chances are we won't shop there. Averages smooth out the fluctuations and extremes of reality, and provide a homogenous substitute. Without question, such averages are useful. However, in some instances we may actually want to focus on, rather than eliminate, the irregularities of a situation. The extremes have the greatest influence on where the average falls, and to overlook them would be to overlook important characteristics of a system.

For example, though we talk about the average occurrence of rainfall, it actually happens in irregular bursts. Rather than dismiss these irregularly spaced bursts, Dr. Shaun Lovejoy of McGill University has studied them closely. He found that the pattern of irregularity holds the same over short and long time spans. Charts for the pattern of rainfall over a period of one year "show, for the most part, the same sort of patterns as rain charts for a month, a week, or a single day" [Econ87].

This scale invariance shows that the distribution of a natural process over time may have a fractal structure. This was also seen by Mandelbrot in the "clump-void" patterns of telephone noise. He made an important discovery when he realized the advantage of using the fractal dimension, rather than the average, as a measure of this intermittence. "The proper measurement of a channel's noisiness," Mandelbrot found, "was not the average number of errors as had seemed obvious, but was a totally unexpected quantity, the Hausdorff-Besicovitch dimension" [Mand82].

Student A
D = 0.4307

Student B
D = 0.8613

Figure 33 Cantor Sets

The clump-void patterns of the Cantor set may also be a useful model to examine a variety of our daily activities. Take homework, for instance. Human concentration is not limitless, so most students do some homework, then none for awhile, and then more later. But within the time they are working, this is broken down into "work, no-work" clumps as well. It looks like "student B" had shorter distractions than "student A" and therefore has a higher fractal dimension for meaningful homework time. Productivity can be measured by the average amount of time spent working; however, measuring the fractal dimension can provide us with some additional information.

As in the case of telephone noise, averages do not explain everything. A fractal model may have "better-than-average" accuracy in analyzing the spread of diseases. Dr. Robert May, a mathematical biologist at Princeton, is working on a prediction based on the theories of chaotic dynamics, concerning the spread of the disease AIDS. In this case, his fractal model is more useful because it recognizes that people behave differently. We are not a homogeneous mass, and therefore "spread the disease at widely varying rates" [Econ87]. As with most natural phenomena, "any model with a fair chance of coming up with right answers has to recognize that society is an irregular cluster" [LaBr87].

Researchers and engineers often make changes to a process with the intent of driving averages up or down. In fractal models, the fractal dimension replaces the average as the prime indicator of the status of a system. At Ohio State University, Katherine Faber uses the fractal dimension of cracks in ceramics to measure the strength and toughness of the materials she uses. The dimensions range from 1.05 to 1.07. Her goal is to make the ceramics stronger and more durable by using secondary materials to bring the dimensions closer to 2 [LaBr87].

Earthquakes and Proteins

Another application of Fractal Geometry lies in the study of earthquakes. In addition to developing fractal theories about the frequencies of earthquakes, seismologists hope to use Fractal Geometry to help predict the destructiveness of the temblors as well. By using the fractal dimension to quantify the jaggedness of a fault, Cathleen Aviles and Chris Scholz at Lamont-Doherty Geological Observatory have found a correlation between the fractal dimension of a fault and the intensity of its seismic behavior. Because dimensions can vary along different regions of a fault, a fractal model can reflect the varying degrees of damage that occur from a single quake.

The fractal dimension may also help scientists in the study of the surfaces of proteins. The surfaces, which in a normal sense would be only two-dimensional, actually exhibit dimensions ranging from 2.4 to 2.7 [LaBr87]. This dimension "varies considerably from place to place along a given protein," [SciN85] and as a result, like "velcro," the bumpier the portion of the protein's surface, the higher its chance of interaction with other surfaces. A categorization of this property using fractal dimension has been used by D.C. Rees of the University of California, Los Angeles, and Mitchell Lewis of SmithKline and French Laboratories in Philadelphia. "Recognition of these geometric factors," they state, "provides a new approach to describing the interaction of macromolecules with one another" [SciN85].

Fractal Aggregates

Researchers have found that fractals are extremely useful in the study of the boundaries between two liquids. In a common "oil-blob-in-water" toy, children can watch colorful "oil-like" liquids cascade through water when the plastic encasement of the toy is flipped or rotated. The "oil," which is thicker, more "syrupy," and what scientists call more viscous than water, tends to stick together in blobs as it travels through the water. Recently, much research has been done to study the patterns of the reverse case. Instead of watching the higher viscosity "oil" dropped through the less viscous "water," scientists are studying the fractal results when a low-viscosity fluid is forced into that of a higher viscosity fluid (i.e. water into oil).

The patterns that develop are a function of the ratio of the viscosities of the two interacting fluids. When oil is injected into glycerin, for example, the surface tension between the two liquids is low enough to allow the spread of the oil, but still high enough to prevent the oil and glycerin from mixing extensively along their borders. The star-shaped pattern that results occurs with a variety of fluids sharing similar ratios of viscosities. The protruding "fingers" of these shapes have earned such patterns the name of "viscous fingering."

Another interesting example occurs when water is injected into clay slurries. In general, the lower the contrast in viscosity between two fluids, the more rounded and undisturbed the fingers are. However, as the contrast between viscosities increases, more "tip splitting" of the fingers occurs, and the resulting shapes become more and more intricate. Eventually they display fractal qualities such as fractal dimension. Scientists are interested in the combination of water and clay because of its implication concerning river deltas.

Such research is also of tremendous interest to oil companies. Oil is contained in porous media and is difficult to extract. Since oil and water are immiscible (not mixable), water can be used to force the oil from reservoirs. As long as a certain force is maintained, the interface between the two liquids remains stable. However, if the boundary between the liquids becomes unstable, the interface becomes more complicated and fingering patterns occur. As the branches, or fingers extend further, oil becomes more easily trapped. "In a typical reservoir where water or gas displaces oil, the same kinds of fingering phenomena occur, leaving much of the oil in the ground" [Wong88]. Such trapping can be a main cause of low oil recovery from reservoirs.

Figure 34 Fractal Aggregate

The figure above is a typical model for the formation of patterns that result when water displaces oil. This model has helped researchers with a bewildering variety of subjects including: the study of dust formation, soot formation, the electrolytic deposit of zinc, the formation of copper clusters, dielectric breakdown such as lightning, the formation of gold particles in colloidal suspension, patterns in alloy films, thinfilm morphology, dendritic solidification, dielectric breakdown, and chemical dissolution.

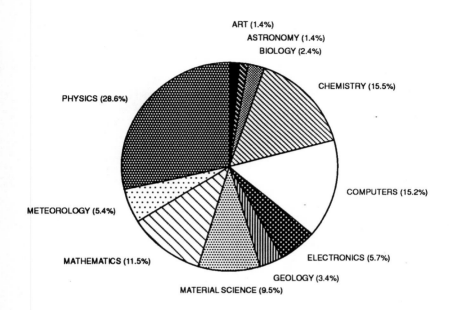

ART (1.4%)
ASTRONOMY (1.4%)
BIOLOGY (2.4%)
CHEMISTRY (15.5%)
PHYSICS (28.6%)
COMPUTERS (15.2%)
METEOROLOGY (5.4%)
MATHEMATICS (11.5%)
ELECTRONICS (5.7%)
GEOLOGY (3.4%)
MATERIAL SCIENCE (9.5%)

Fractal Research

The above breakdown of fractal research comes from a survey of 179 fractal-related articles published from 1985-1988. The fact that the physics, chemistry, and computer-related applications outnumber purely mathematical topics indicates that the "applied fractal" has indeed arrived. The scope of the literature ranges from popular science magazines to the highly technical, and specialized academic journals. A complete listing of subjects covered in the articles follows, and though the list is extensive, the sampling of articles from which it is derived is estimated to be just 20% of available reading on Fractal Geometry and its applications. The applications listed, therefore, are by no means exhaustive.

FRACTAL APPLICATIONS
(a partial list)

acid rain
adsorption
aggregation
alkanoic acids
alloy films, patterns
anistrophy
art
arthropods
astronomy
atmospheric boundary layer
attractors
beams and girders, vibration
botany
brownian movements
cache-memory devices
catastrophe theory, mathematics
ceramic materials
ceramic powders
chaos
chaotic behavior in systems
chemical dissolution
chemical reactions, computer simulations
chemical reactions, kinetics
chemistry
city planning
clay
clouds
cluster set theory
coagulation
colloids
computer graphics
computer graphics, artists and designers use
computer graphics, surface and contour representation
computers
computers, simulation programs
conformational analysis
cosmology
cray computers
crystallization
data compression
data structures (computer science)
dendritic solidification
deterministic chaos
dielectrics
dielectric breakdown
dielectric constants
diffraction
diffraction, mathematical models

diffusion
diffusion, computer simulation
digital techniques
dipole moment
dust formation
dyes, spectra
dynamics
earthquakes, fault patterns
earthquakes, frequency of
ecosystems, computer simulation
electric resistance, mathematical models
electro-chemistry
electrodeposition of metals
electrodeposition of metals, mathematical models
electrodes, design
electromagnetic scattering
electron energy
electronic circuits
energy levels
energy transfer
entrainment (meteorology)
entropy
epidemics
exchange rates, fluctuation of
finite element method
flames, premixed flames
flames, turbulent flames
flocculation
fluids
fluids, two-phase flow
fluorescence, quenching
formation of copper clusters
fracture strength
Fresnel diffraction
galaxies, distribution of
geology
geology, stratigraphic, mathematical models
geology, statistical methods
graphs, large networks
heat conductivity
Hilbert space
hydrodynamics
image compression
image processing
image processing resolving power (optics)
impedance
information theory
insulation (electric), failure of
iterated function systems

jet mixing
kinetics
landscapes, distribution of vegetation
LANDSAT satellites
laser beams
light scattering
lightning, dielectric breakdown
Mandelbrot set
material science
mathematical models
measurement
measurement errors
medical research
metal oxide semiconductors
metals
metals, failure of
metallurgy
meteorological optics
meteorology
meteorological stations
microemulsions
molecules
Monte-Carlo methods
multiprocessors (computers), design of
music, computer generated
napthalene
networks
nickel alloys, metallography
numbers, complex
oil extraction
optical coatings, manufacture of
optical properties
optics
optics, geometrical
ore grade vs. tonnage
ore reserves
oscillators
particle size
particles, instant coffee
percolation
phased arrays
phonons
phospholipids
photochemistry
physical chemistry
physiology
pitch (sound)
plates, vibration of
polymers
population biology
porous materials
probabilities
proteins
proteins, analysis of
quantum mechanics
radio antennas, mathematical models

recombination reactions, computer
    simulation
refractive index
relativity
rocks, sedimentary
scattering, ratio and proportion
self-diffusion (solids)
self-organized systems
set theory
silver compound, spectra
snow crystals
solar batteries
solitons
soot, formation/patterns of
sound scattering
species flux
spectrum, intensity
statistical mechanics
steel, cleaning of
steel, failure of
stochastic processes
strength of materials
structural engineering
structural engineering/finite element
    method
surface and contour representation
surface chemistry
surface phenomena
surface roughness
surface roughness, measurement of
surfaces/areas and volumes
symmetry
thermal transport
thermography
thinfilm morphology
thin metallic films
turbulence
universe, structure of
vibrations
viscosity
viscous fingering
weather forecasting
x-rays, scattering of
zeolites

CONCLUSION

"It is pleasing that fractals often lead to attractive graphics that help in identifying and handling their applications. It is even more pleasing that they should be related to active areas of analytic physics, and rooted in old and admirable mainstream mathematics."

[Mand78]

THE PROMISE OF GEOMETRY

Mathematics of the 21st century offers a new frontier. This frontier is marked by stunning images; some entirely artificial, and others that model nature with ease and remarkable likeness. Its founders embraced rather than rejected a geometry of nature, and though still in its infancy, the field of Fractal Geometry has attracted an exceptionally broad audience. Educators, hobbyists, and children can explore and even create their own fractals, and a child as well as a leading scientist can study worlds of untouched beauty with a renewed interest in visual exploration.

In the past, mathematicians' tools for geometrical description were largely confined to a small "Euclidean" alphabet. Though the field is still in its infancy, the theory of Fractal Geometry has been established, and these shapes with their fractal dimension and self-similarity are very often nature's rule, as opposed to its exception. Historically, geometry "had to ignore the crinkles, whorls, squiggles, and billows of the real world because they did not submit to standard mathematical formulae" [Econ87]. Now, equipped with an adequate theoretical as well as experimental base, Fractal Geometry promises to rival Euclid.

"Now, 12 years after Dr. Mandelbrot wrote his book, The Fractal Geometry of Nature, the evidence that fractals can shed light on a wide variety of problems is piling up. The applied fractal has arrived. Armed with a technique of measuring the irregularity of shapes, the theory of fractals has now been applied to protein structure, acid rain, earthquakes, fluctuation of exchange rates, oil extraction, epidemics, corrosion, brittle fractures, music, distribution of galaxies, levels of the Nile, shapes of clouds, trees, lakes and mountains. Nearly every branch of science studies something that fractals can help with, because all aspects of nature involve some roughness or irregularity."

[Econ87]

The ability to recognize shapes as fractal lies in fine-tuning our ability to recognize self-similarity and fractal dimension in the world around us. An ability to point out this fundamental contrast to Euclidean geometry is central to understanding Fractal Geometry. As Mandelbrot summed up so beautifully:

"Clouds are not spheres,
mountains are not cones,
trees are not cylinders,
bark is not smooth,
and neither does lightning travel in a straight line."

[Mand82]

Fractal Geometry succeeds in highlighting the beauty and applicability of mathematics within a continually expanding list of fields. But as Mandelbrot suggests, perhaps fractals' "most obvious and indisputable usefulness has been to bring geometry back to the source indicated by the etymology of 'geometry': the measurement of the Earth" [Mand83].

APPENDICES

The "dragon curve", also known as the "self-squared dragon," is one of the most basic fractals. It falls under the class of fractals described as self-replicating, or linear. A basic shape is produced, generally using straight lines that have an orientation indicated by an arrow. Each line will in turn be replaced by the self-replicated image of the original pattern. This is repeated on each progressing level. As the level of replication continues, a pattern emerges. If the pattern will eventually fill a region completely, as does the dragon curve, it is considered 2-dimensional. If it does not, it has a lower dimension.†

With only two lines of information the proceeding dragon curve emerges.

Places initial object upward.   Places initial object downward.

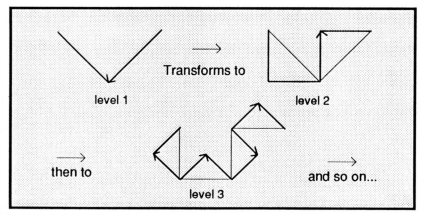

† Chapter 3 discusses linear dimensions.

Transformation of the dragon curve

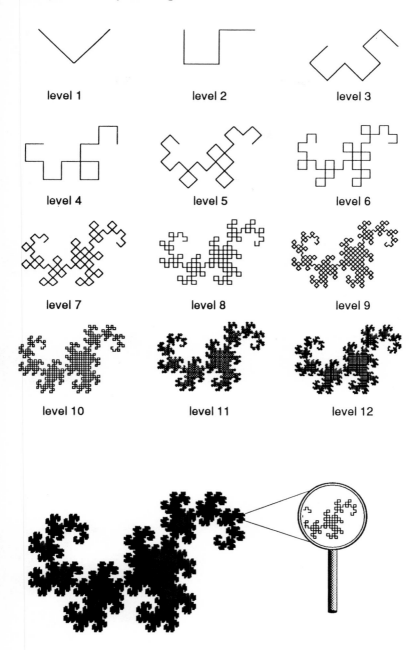

level 1

level 2

level 3

level 4

level 5

level 6

level 7

level 8

level 9

level 10

level 11

level 12

Zoom of a high level dragon curve.

In the fractal world, simple shapes with simple rules can create very complex-looking objects. The case of the dragon curve is a good example. Let's explore it further.

As it replicates itself, each line of the initial seed can be oriented in four different directions: upward from left to right, upward from right to left, downward from left to right, and downward from right to left. Since there are two line segments in the seed, with four possible orientations each, there are a total of sixteen possibilities. What would happen if we changed the orientation of the two line segments in this basic example?

There are certain things we can predict about our curves. The same seed will create an area-filling, two-dimensional picture, regardless of which way the fractal's lines are pointing. We start with similar line lengths, and each pattern is iterated for 10 generations (2^{10} line segments resulting from seed replacement 10 times). Since the seeds are mirror images of one another, their outcomes reflect this. On the opposite page you can see the results. Since some of the seeds produce duplicate patterns, they are shown only once. Below are the seeds:

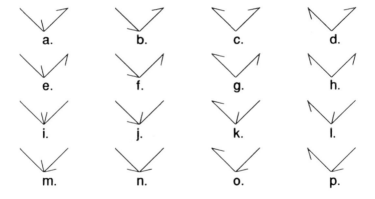

Try to match the seeds with their resulting transformations.

Helpful hints: Using graph paper starting a seed squares (32 x 16) can be most useful. After just a few replacement levels you should start seeing a pattern emerging.

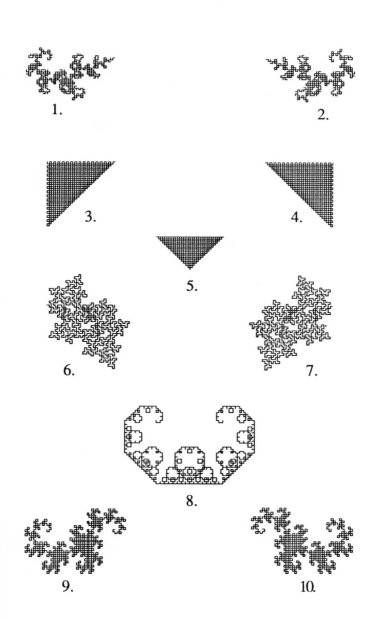

1.

2.

3.

4.

5.

6.

7.

8.

9.

10.

There are a number of objects below, and each has been generated using a number of 1-cm line segments. Then the segments have been replaced with a small replica of the whole object. This was repeated 4 times (level 4). The goal of the game is to put the objects in order of their fractal dimension, and to find their underlining seed-shape. Hints to finding solutions are: 1) divide the object into halves or thirds, 2) look at the ends for a recognizable seed-shape of the whole, 3) check the density of area being covered. Generally, the more solid the area is, the higher the fractal dimension. Also notice disproportionately long lines, as they generally indicate lines that have not been transformed.

ANSWERS ON NEXT PAGE

Answerers to Fractal Totem Pole:

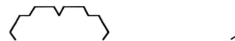

Cantor set 0.6309D, 2-1cm line segments, level 4

Straight line 1.0000D, 3-1cm line segments, level 4

Bulging pyramid 1.2618, 2-1cm line segments, level 4

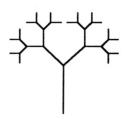

Koch curve 1.2619D, 4-1cm line segments, level 4

Tree 1.3041D, 3-1cm line segments, level 4

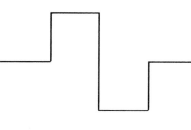

Koch-blocks 1.5000D, 8-1cm line segments, level 4

 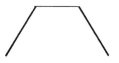

Sierpinski triangle 1.5850D, 3-1cm line segments, level 4

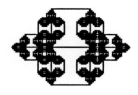 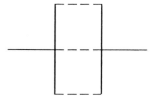

Quasi-Peano curve 1.6309D, 9-1cm line segments, level 4

 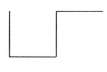

Dragon curve 2.0000D, 4-1cm line segments, level 4

 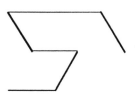

Peano-Gosper 2.0000D, 7-1cm line segments, level 4

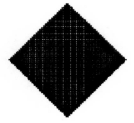 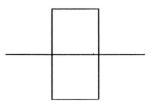

Peano curve 2.0000D, 9-1cm line segments, level 4

FOOTNOTES

[Barn88]	Barnsley, M. *Fractals Everywhere*. San Diego: Academic Press, Inc. 1988.
[Batt85]	Batty, M. "Fractals — Geometry Between Dimensions." New Scientist 105. 4/4/85 (1985): 31-5.
[deMi90]	de Miranda, C. 1990
[Econ87]	Editors. "Tomorrow's Shapes: The Practical Fractal." Economist. Dec. 26, '87 (1987): 99-103.
[Frae86]	Fraedrich, K. "Estimating the Dimensions of Weather and Climate Attractors." Journal of the Atmospheric Sciences 43. 3/1/86 (1986): 419-32.
[Garc88]	Garcia, L. "Fractal Geometry." 1988. Unpublished.
[Glei88]	Gleick, J. *Chaos*. New York: Viking Press,1988.
[Hort88]	Hortmann, M. Visiting Professor UCSC, 1987/88
[Jurg90]	Jurgens, H., Peitgen, H.O., Saupe, D. "The Language of Fractals." Scientific American. 8/90. (1990): 60-67.
[Kozl86]	Kozlov, A. "Hopping Heat Waves: The Arcane Geometry of Fractals Finds Order in Chaos." Science Digest 94. Jun. 1987 (1986): 28.
[LaBr87]	La Brecque, M. "Fractal Applications." Mosaic 17.4. Winter 1986/87 (1986/87): 34-48.
[Lakh88]	Lakhtakia, A. "The Bohr-Hund Atom is a Fractal!" American Journal of Physics 56. 2/88 (1988): 104-5.
[Mand78]	Mandelbrot, B. "Getting Snowflakes into Shape." New Scientist. 6/22/78 (1978): 808-810.
[Mand82]	Mandelbrot, B. *The Fractal Geometry of Nature*. San Francisco: W.H. Freeman and Company, 1982.
[Mand83]	Mandelbrot, B. "On Fractal Geometry, and a Few of the Mathematical Questions It Has Raised." Proceedings of the International Congress of Mathematicians 1, 2. August 16-24, 1983 (1983): 1661-1675.

[Meak88]   Meakin, P., Stanley, H.E. "Multifractal Phenomena in Physics and Chemistry." Nature 335. 9/29/88 (1988): 405-9.

[MInt85]   Editors. "Mandelbrot Recognized for Major Scientific Achievement." Mathematical Intelligencer 7 no 4. 1985 (1985): 64.

[Peit86]   Peitgen, H. O. , Richter, P. H. *The Beauty of Fractals.* Heidelberg: Springer-Verlag, 1986.

[Scha89]   Schaefer, D. "Polymers, Fractals and Ceramic Materials." Science 243. 2/24/89 (1989): 1023-7.

[SciN85]   Editors. "The Texture of Clinging Proteins." Science News 128. 12/14/85 (1985): 377.

[Thom87]   Thomsen, D. E. "Fractals: Magical Fun or Revolutionary Science?" Science News 131. Mar. 21, '87 (1987): 184.

[Turc86]   Turcotte, D. "A Fractal Approach to the Relationship Between Ore Grade and Tonnage." Economic Geologists and the Bulletin of Economic Geologists 81. 9,10/86 (1986): 1528-32.

[Whit89]   Whitby, J. "Teaching Fractals in High School." 1989. Unpublished.

[Zorp88]   Zorpette, G. "Fractals: Not Just Another Pretty Picture." IEEE Spectrum 25. 10/88 (1988): 29-31.

Ackerman, M. "Hilbert curves made simple." Byte 11. 6/86 (1986): 137-8.

Anacker, L., Parson, R., Kopelman, R. "Diffusion-controlled reaction kinetics on fractal and Euclidean lattices: transient and steady-state annihilation." Journal of Physical Chemistry 89. 10/24/85 (1985): 4758-61.

Anacker, L. Kopelman, R. "Steady-state chemical kinetics on fractals: geminate and nongeminate generation of reactants." Journal of Physical Chemistry 91. 10/22/87 (1987): 5555-7.

Avnir, David. "The geometry factor in photoprocesses on irregular surfaces." Journal of the American Chemical Society 109. 5/13/87 (1987): 2931-8.

Bak, Per. "The devils staircase." Physics Today 39. 12/86 (1986): 38-45.

Bak, P., Tang, C. "Self-Organized criticality." Physics Today 42. 1/89 (1989): S27-8.

Banchoff, Thomas F. "Book Review: The Beauty of Fractals." American Scientist 76. 11,12/88 (1988): 606-7.

Barcellos, Anthony. "The Fractal Geometry of Mandelbrot." College Mathematics Journal 15. 3/84 (1984): 98-114.

Barnsley, M., Sloan, A., and . "A better way to compress images." Byte 13. 1/88 (1988): 215-18+.

Barnsley, M. Fractals Everywhere. San Diego: Academic Press, Inc. 1988.

Batty, Michael. "Fractals— geometry between dimensions." New Scientist 105. 4/4/85 (1985): 31-5.

Bedrosian, S., Jaggard, D. "A fractal-graph approach to large networks." Proceeding of the IEEE 75. 7/87 (1987): 966-8.

Benzoni, J., Sarkar, S., Sherrington, D. "Random phase screens." Journal of the Optical Society of America; Optics and Image Science 4. 1/87 (1987): 17-26.

Berry, M.V. "Disruption of images: the caustic-touching theorem." Journal of the Optical Society of America; Optics and Image Science 4. 3/87 (1987): 561-9.

Berry, Michael. "Book Review: The Science of fractal images." New Scientist 121. 3/11/89 (1989): 66.

Bobinsky, Eric. "Book Review: The Science of Fractal Images." Byte 13. 12/88 (1988): 51.

Brady, Ball. "Fractal growth of copper electrodeposits." Nature 309. 5/17/84 (1984): 225-9.

Bridger, Mark. "Looking at the Mandelbrot set." College Mathematics Journal 19. 9/88 (1988): 353-63.

Broggi, G. "Evaluation of dimensions and entropies of chaotic systems." Journal of the Optical Society of America: Optical Physics 5. 5/88 (1988): 1020-8.

Cahalan, R., Joseph, J.H. "Fractal statistics of cloud fields." Monthly Weather Review 117. 2/89 (1989): 261-72.

Cannata, F., Ferrari, L. "Dimensions of relativistic quantum mechanical paths." American Journal of Physics 56. 8/88 (1988): 721-5.

Church, Eugene L. "Fractal surface finish." Applied Optics 27. 4/15/88 (1988): 88.

Ciarcia, Steve. "A supercomputer (Circuit Cellar Mandelbrot engine)." Byte 13. 10/88 (1988): 283-6.

Cipra, Barry. "Computer-drawn Pictures Stalk the Wild Trajectory." Science 241. Sep. 2, '88 (1988): 1162-3.

Cipra, Barry A. "Image capture by computer." Science 243. 3/10/89 (1989): 1288-9.

Clark, Nigel N. "Fractal harmonies and rugged materials." Nature 319. 2/20/86 (1986): 625.

Cohen, R.D. "Effect of interaction energy on floc structure." AICHE Journal 33. 9/87 (1987): 1571-5.

Courtens, Eric. "Observations of fractons." Physics Today 42. 1/89 (1989): S31.

Crutchfield, J., Farmer, J., Packard, N.H. "Chaos." Scientific American 255. 12/86 (1986): 46-57.

Daccord, Lenormand. "Fractal patterns from chemical dissolution." Nature 325. 1/1/87 (1987): 41-3.

Davidson, J., Keller, D. "The measurement of image sharpness through the approaches used to describe fractals." SMPTE Journal 94. 8/85 (1985): 802-9.

Devaney, Robert L. "Chaotic Bursts in Nonlinear Dynamical Systems." Science 235. Jan. 16, '87 (1987): 342-5.

Dewdney, A. K. "Probing the Strange Attractions of Chaos." Scientific American 257.

Dewdney, A.K. "Of fractal mountains, graftal plants and other computer graphics at Pixar." Scientific Amercican 255. 12/86 (1986): 14-16+.

Dewdney, A.K. "Beauty and profundity: the Mandelbrot set and a flock of its cousins called Julia." Scientific American 257. 11/8 (1987): 140-5.

Dewdney, A.K. "Random walks that lead to fractal crowds." Scientific American 259. 12/88 (1988): 116-19.

Dewdney, A.K. "Computer recreations; a tour of the Mandelbrot set aboard the Mandelbus." Scientific American 260. 2/89 (1989): 108-11.

Dodge, C., Bahn, C. "Musical fractals." Byte 11. 6/86 (1986): 185+.

Dowell, E. Pezeshki, C. "Book Review: Chaotic vibrations; an intro to chaotic dynamics for applied scientists and engineers." Journal of Applied Mechanics 55. 12/88 (1988): 997-8.

Editors. "Additional perspectives of fractals." College Mathematics Journal 15. 3/84 (1984): 115-19.

Editors. "Fractals: aggregationism." Nature 309. 5/17/84 (1984): 214.

Editors. "Mathematics of a liquid squeeze play." Science 85. 6/85 (1985): 9.

Editors. "The texture of clinging proteins." Science News 128. 12/14/85 (1985): 377.

Editors. "Interfaces groove fractally." Science News 128. 8/17/85 (1985): 105.

Editors. "Mandelbrot recognized for major scientific achievement." Mathematical Intelligencer 7 no 4. 1985 (1985): 64.

Editors. "Fractal Geometry (computer graphics)." Electronics and Power 32. 8/86 (1986): 581.

Editors. "Recursive geography." Electronics and Wireless World 92. 6/86 (1986): 5.

Editors. "A fractal universe?." Science News 129. 4/5/86(1986): 217.

Editors. "How many dimensions?." Sky and Telescope 72. 8/86 (1986): 125-6.

Editors. "Unmashing chaos." Scientific Amercican 256. 2/87 (1987): 61-2.

Editors. "Tomorrow's Shapes: The Practical Fractal." Economist. Dec. 26, '87 (1987): 99-103.

Evesque, P., Yang, C., El-Sayed, M. "Comparison between electrodeposited aggregates in two dimensions and the fractal pattern calculated by the Witten-Sander model." Journal of Physical Chemistry 90. 5/22/86 (1986): 2519-21.

Farin, Volpert, Avnir. "Determination of adsorption conformation from surface resolution analysis." Journal of the American Chemical Society 107. 5/29/85 (1985): 3368-70.

Farin, D., Avnir, D. "Reactive fractal surfaces." Journal of Physical Chemistry 91. 10/22/87 (1987): 5517-21.

Fraedrich, Klaus. "Estimating the dimensions of weather and climate attractors." Journal of the Atmospheric Sciences 43. 3/1/86 (1986): 419-32.

Gleick, J. Chaos. New York: Viking Press,1988.

Gomes, M.A.F. "Fractal geometry in crumpled paper balls." American Journal of Physics 55. 7/87 (1987): 649-50.

Gouldin, F.C. "Interpretation of jet mixing using fractals." AIAA Journal 26. 11/88 (1988): 1405-7.

Grebogi, Ott, Yorke. "Chaos, strange attractors, and fractal basin boundaries in nonlinear dynamics." Science 238. Oct. 87 (1987): 632-8.

Gregogi, Ott, Yorke. "Chaos, strange attractors, and fractal basin boundaries in non-linear dynamics." Science 238. 10/30/87 (1987): 632-8.

Guenther, Karl H. "Growth structures in a thick vapor deposited MgF2 multiple layer coating." Applied Optics 26. 1/15/87 (1987): 188-90.

Hawton, M., Keeler, W. "Anomalous low-frequency dispersion in phospholipid-water multilayers." Journal of Applied Physics 64 pt1. 11/15/88 (1988): 5088-91.

Hepel, Tadeusz. "Effect of surface diffusion in electrodeposition of fractal structures." Journal of the Electro-Chemical Society 134. 11/87 (1987): 2685-90.

Herbert, Franz. "Fractal landscape modeling using octrees." IEEE Computer Graphics and Applications 4. 11/84 (1984): 4-5.

Hmurcik, L., Serway, R.A. "Frequency dispersion in the admittance of the polycrystalline Cu2S/CdS solar cell." Journal of Applied Physics 61. 1/15/87 (1987): 756-61.

Holdun, Arun. "Chaos Is No Longer A Dirty Word." Scientist 106. Apr. 25, '85 (1985): 12-15.

Hollingsworth, A. "Weather forecasting: storm hunting with fractals." Nature 319. 1/2/86 (1986): 11-12.

Horgan, John. "Fractal shorthand." Scientific Amercican 258. 2/88 (1988): 28.

Hoshen, J., Kopelman, R., Newhouse, J. "Refined Monte-Carlo simulations of static percolation." Journal of Physical Chemistry 91. 1/1/87 (1987): 219-22.

Huang, L., Ding, J., Li, H. "Growth of the fractal patterns in Ni-Zr thin films during ion-solid interaction." Journal of Applied Physics 63 pt 1. 4/15/88 (1988): 2879-81.

Hughes, Ferencz, Hallquist. "Large-scale vectoried implicit calculation in solid mechanics on a Cray X-Mp/48 utilizing EBE precondition conjugate gradients." Computer Methods in Applied Mechanics and Engineering 61. 3/87 (1987): 215-48.

Hurd, Alan J. "Resource letter FR-1: fractals." American Journal of Physics 56. 11/88 (1988): 969-75.

Jaggard, D., Kim, Y. "Diffraction by band-limited fractal screens." Journal of the Optical Society of America; Optics and Image Science 4. 6/87 (1987): 1055-62.

Jakeman, E., Renshaw, E. "Correlated random-walk model for scattering." Journal of the Optical Society of America; Optics and Image Science 4. 7/87 (1987): 1206-12.

Jeffery, Tom, and . "Mimicking Mountains." Byte 12. 12/87 (1987): 337-8+.

Jeng, J., Varaden, V.V., Varaden, V.K. "Fractal finite element mesh generation for vibration problems." Journal of the Acoustical Society of America 82. 11/87 (1987): 1829-33.

Jullien, R. "Aggregation phenomena and fractal aggregates." Contemporary Physics 28. 9,10/87 (1987): 477-93.

Kadanoff, Leo P. "Fractals: Where's the Physics?" Physics Today 39. Feb. '86 (1986): 6-7.

Kamada, Ray F. "A fractal model for dry convective boundary layers." Journal of the Atmospheric Sciences 45. 9/1/88 (1988): 2365-83.

Kandebo, Stanley W. "Fractals research furthers digitized image compression." Aviation Week & Space Technology 128. 4/25/88 (1988): 91+.

Kerstein, Alan R. "Fractal dimension of turbulent premixed flames." Combustion Science and Technology 60 no 4-6. 1988 (1988): 441-5.

Kim, Y., Jaggard, D.L. "The fractal random array." Proceeding of the IEEE 74. 9/86 (1986): 1278-80.

Kim, Y., Jaggard, D. "Optical beam propagation in a band-limited fractal medium." Journal of the Optical Society of America; Optics and Image Science 5. 9/88 (1988): 1419-26.

Kim, Y., Jaggard, D.L. "Band-limited fractal model of atmospheric refractivity fluctuation." Journal of the Optical Society of America; Optics and Image Science 5. 4/88 (1988): 475-80.

Kopelman, Raoul. "Fractal reaction kinetics." Science 241. 9/23/88 (1988): 1620-6.

Kozlov, Alex. "Hopping Heat Waves: The Arcane Geometry of Fractals Finds Order in Chaos." Science Digest 94. Jun. 1987 (1986): 28.

Kuo, K., Welch, R., Sengupta, S. "Structural and textural characteristics of cirrus clouds observed using high spatial resolution LANDSAT imagery ." Journal of Applied Meteorology 27. 11/88 (1988): 1242-60.

La Brecque, Mort. "Fractal Applications." Mosaic 17.4. Winter 1986/87 (1986/87): 34-48.

La Brecque, Mort. "Fractals in Physics." Mosaic 18. Sum 87 (1987): 22-41.

Lakhtakia, A., Varadan, V.K., Varadan, V.V. "Scalar scattering characteristics of a periodic, impenetrable surface: effect of surface modeling errors." Journal of Applied Physics 60. 12/15/86 (1986): 4090-4.

Lakhtakia, Akhlesh. "The Bohr-Hund atom is a fractal!" American Journal of Physics 56. 2/88 (1988): 104-5.

Land, Bruce R. "Dragon (dragon curve on the Macintosh)." Byte 11. 4/86 (1986): 137-8.

Levi, Barbara G. "New global fractal formalism describes paths to turbulence." Physics Today 39. 4/86 (1986): 17-18.

Lewis, M., Rees, D.C. "Fractal surfaces of proteins." Science 230. 12/6/85 (1985): 1163-5.

Lianos, P., Modes, S. "Fractal modeling of luminescence quenching in microemulsions." Journal of Physical Chemistry 91. 11/19/87 (1987): 6088-9.

Ligon, Woodfin V. "Fractal structures obtained by electrodeposition of Ag." Journal of Chemical Education 64. 12/87 (1987): 1053.

Ling, Frederick F. "Scaling law for contoured length of engineering surfaces." Journal of Applied Physics 62. 9/15/87 (1987): 2570-2.

Lovejoy, Schertzer, Ladoy. "Fractal characterization of inhomogeneous geophysical measuring networks." Nature 319. 1/2/86 (1986): 43-4.

Maddox, John. "Aggregation by very large numbers." Nature 318. 11/21/85 (1985): 229.

Maddox, John. "Gentle warning on fractal fashions." Nature 322. 7/24/86 (1986): 303.

Maddox, John. "The universe as a fractal structure." Nature 329. 9/17/87 (1987): 195.

Mandelbrot, B. "Getting snowflakes into shape." New Scientist. 6/22/78 (1978): 808-810.

Mandelbrot, Benoit B. "On Fractal Geometry, and a Few of the Mathematical Questions It Has Raised." Proceedings of the International Congress of Mathematicians 1, 2. August 16-24, 1983 (1983): 1661-1675.

Mandelbrot, Passoja, Paullay. "Fractal character of fracture surfaces of metals." Nature 308. 4/19/84 (1984): 721-2.

Mandelbrot, B. The Fractal Geometry of Nature. San Francisco: W.H. Freeman and Company, 1982.

Martin, J., Bentz, D. "Fractal based description of the roughness of blasted steel panels." Journal of Coatings Technology 59. 2/87 (1987): 35-41.

McDermott, Jeanne. "Dancing to fractal time." Technology Review 92. 1/89 (1989): 6+.

McWorter, W., Tazelaar, J., and . "Creating fractals." Byte 12. 8/87 (1987): 123-8+.

Mello, Tufillaro. "Strange attractors of a bouncing ball." American Journal of Physics 55. 4/87 (1987): 316-20.

Meyer, Farin, Avnir. "Cross-sectional areas of alkanoic acids. Application of fractal theory of adsorption and consideration of molecular shape." Journal of the American Chemical Society 108. 12/10/86 (1986): 7897-905.

Milder, D., Hall, P. "Small-angle scattering from porous solids with fractal geometry." Journal of Physics, Applied Physics 19. 8/14/86 (1986): 1535-45.

Morse, Lawton, Dodson. "Fractal dimension of vegetation and the distribution of arthropod body lengths." Nature 314. 4/25/85 (1985): 731-3.

Mu, Z., Lung, C.W. "Studies on the fractal dimension and fracture toughness of steel." Journal of Physics 21. 5/14/88 (1988): 848-50.

Musgrave, F.K., Mandelbrot, B. "Natura ex machina (fractal landscape images)." IEEE Computer Graphics and Applications 9. 1/89 (1989): 4-7.

Nauenberg, Michael. "Book Review: Aggregation & Fractal Aggregates." American Scientist 76. 9,10/88 (1988): 508.

Neal, Margaret. "The visual think (fractals in graphic art)." IEEE Computer Graphics and Applications 8. 1/88 (1988): 3-5.

Niklasson, G. A. "Fractal aspects of the dielectric response of charge carriers in disordered materials." Journal of Applied Physics 62 . 10/1/87 (1987): R1-14.

Niklasson, G.A. "Optical properties of gas-evaporated metal particles: effects of a fractal structure." Journal of Applied Physics 62. 7/1/87 (1987): 258-65.

Nittmann, Daccord. "Fractal growth of viscous fingers: quantitative characterization of a fluid instability phenomenon." Nature 314. 3/14/85 (1985): 141-4 .

Nittmann, Stanley. "Tip splitting without interfacial tension and dendritic growth patterns arising from molecular anisotropy." Nature 321. 6/12/86 (1986): 663-8.

Onodo, G, Toner, J. "Fractal dimensions of model particle packings having multiple generations of agglomerates." Journal of the American Ceramic Society 69. 11/86 (1986): C278-9.

Orbach, Roy. "Dynamics of fractal networks." Science 231. 2/21/86 (1986): 814-19.

Ornstein, D. S. "Ergodic Theory, Randomness, and Chaos." Science 243. Jan. 3, '89 (1989): 182-7.

Oxaal, Murat, Boger. "Viscous fingering on percolation clusters." Nature 329. 9/3/87 (1987): 32-7.

Pajkossy, T., Nyikos, L. "Impedance of fractal blocking electrodes." Journal of the Electro-Chemical Society 133. 10/86 (1986): 2061-4.

Peitgen, H. O. , Richter, P. H. The Beauty of Fractals. Heidelberg: Springer-Verlag, 1986.

Peitgen, H.O., Saupe, D. The Science of Fractal Images. New York: Springer-Verlag, 1988.

Peleg, Normand. "Characterization of the ruggedness of instant coffee particle shape by natural fractals." Journal of Food Science 50. 5,6/85 (1985): 829-31.

Pendry, John. "Symmetry and transport in disordered systems." IBM Journal of Research and Development 32. 1/88 (1988): 137-43.

Peterson, Ivars. "Fractals For Modeling Ecosystems." Science News 123. Jun. 11, '8f3 (1983):

Peterson, Ivars. The Mathematical Tourist. New York: W.H. Freeman and Company, 1988.

Peterson, Ivars. "Packing It In: Fractals Play an Important Role In Image compression." Science News 131. May 2, '87 (1987): 283-5.

Peterson, Ivar. "Zeroing In On Chaos (with Newton's method)." Science News 131. Feb. 28, '87 (1987): 137-9.

Pines, D., Huppert, D. "Time resolved fluorescence depolarization measurement in porous silicas. The fractal approach." Journal of Physical Chemistry 91. 12/31/87 (1987): 6569-72.

Plotnick, Roy E. "A fractal model for the distribution of stratigraphic hiatuses." Journal of Geology 94. 11/86 (1986): 885-90.

Poddar, B., Moon, F., Mukherjee, S. "Chaotic motion of an elastic-plastic beam." Journal of Applied Mechanics. 55. 3/88 (1988): 185-9.

Prasad, J., Kopelman, R. "Fractal-like molecular reaction kinetics: solute photochemistry in porous membranes." Journal of Physical Chemistry 91. 1/15/87 (1987): 265-6.

Prusinkiewicz, P., Sandness, G. "Koch curves as attractors and repellers." IEEE Computer Graphics and Applications 8. 11/88 (1988): 26-40.

Robinson, Arthur L. "Fractal fingers in viscous fluids." Science 228. 5/31/85 (1985): 1077-80.

Rudnick, Gaspari . "The shapes of random walks." Science 237. 7/24/87 (1987): 384-9.

Sander, L.M. "Scale-invariant phenomena: viscous fingers and fractal growth." Nature 314. 4/4/85 (1985): 405-6.

Sander, L.M. "Fractal growth processes." Physics Today 38. 1/85 (1985): S19.

Sander, Leonard M. "Fractal growth processes." Nature 322. 8/28/86 (1986): 789-93.

Sander, Leonard M. "Fractal growth." Scientific Amercican 256. 1/87 (1987): 94-100.

Schaefer, Dale W. "Fractals and small-angle scattering." Physics Today 40. 1/87 (1987): S21-S22.

Schaefer, Dale W. "Polymers, fractals and ceramic materials." Science 243. 2/24/89 (1989): 1023-7.

Schroeder, Peter B. "Plotting the Mandelbrot Set." Byte 11. 12/86 (1986): 207-10.

Schroeder, Manfred R. "Auditory paradox based on fractal waveform." Journal of the Acoustical Society of America 79. 1/86 (1986): 186-9.

Siiman, O., Feilchenfeld, H. "Internal fractal structure of aggregates of silver particles and its consequences on surface-enhanced Raman scattering intensities." Journal of Physical Chemistry 92. 1/28/88 (1988): 453-64.

Simons, S. "The effect of coagulation on steady-state Brownian diffusion for particles with a fractal structure." Journal of Physics 20. 9/14/87 (1987): 1197-9.

Smalley, R.F., Chatelain, J., Turcotte, D. "A fractal approach to the clustering of earthquakes: application to the seismicity of the New Hebrides." Bulletin of the Seismological Society of America 77. 8/87 (1987): 1368-81.

Stanley, H.E. , Meakin, P. "Multifractal phenomena in physics and chemistry." Nature 335. 9/29/88 (1988): 405-9.

Stewart, Ian. "Les chroniques de rose polymath: les fractals (cartoon giving children a trip into the 2.5th dimension)." Mosaic 16. 1,2/85 (1985): 10-13.

Stewart, Ian. "Classical continued fractals." Nature 318. 12/12/85 (1985): 512.

Stewart, Ian. "The beat of a fractal drum." Nature 333. 5/19/88 (1988): 206-7.

Sutton, Christine. "How fractals help physicists to make snowflakes." New Scientist 105. 4/4/85 (1985): 32-3.

Taubes, Gary. "The Mathematics of Chaos." Discover . Sep. 84 (1984):

Termonia, Meakin. "Formation of fractal cracks in a kinetic fracture model." Nature 320. 4/3/86 (1986): 429-31.

Thibaut, Dominique. "From the fractal dimension of the intermiss gaps to the cache-miss ratio." IBM Journal of Research and Development 32. 11/88 (1988): 796-803.

Thomsen, Dietrick E. "Making Music - Fractally." Science News 117. Mar. 22, '80 (1980): 187, 190.

Thomsen, Dietrick E. "A Place In The Sun For Fractals." Science News 121. Jan. 9, '82 (1982): 28, 30.

Thomsen, Dietrick E. "Fractals: magical fun or revolutionary science?" Science News 131. Mar. 21, '87 (1987): 184.

Till, Johns. "Fractal-based algorithm gets 1.25 bit/pixel compression." Electronic Design 36. 8/11/88 (1988): 34+.

Turcotte, Smalley, Solla. "Collapse of loaded fractal trees." Nature 313. 2/21/85 (1985): 671-2.

Turcotte, D. . "A fractal approach to the relationship between ore grade and tonnage." Economic Geologists and the Bulletin of Economic Geologists 81. 9,10/86 (1986): 1528-32.

Van Damme, Obrecht, Levitz. "Fractal viscous fingering in clay slurries." Nature 320. 4/24/86 (1986): 731-3.

Wagner, Colvin, Trevor, James. "Fractal models of protein structure, dynamics, and magnetic relaxation." Journal of the American Chemical Society 107. 10/2/85 (1985): 5589-94.

Walker, Jearl. "Fluid interfaces, including fractal flows, can be studied in a Hele-Shaw cell." Scientific Amercican 257. 11/87 (1987): 134-6+.

Waters, Tom. "Fractals in your future." Discover 10. 3/89 (1989): 26+.

Weisburd, Stefi. "Fractals, fractures and faults." Science News 127. 5/4/85 (1985): 279.

Weismann, H.J., Zeller, H.R. "A fractal model of dielectric breakdown and prebreakdown in solid dielectrics." Journal of Applied Physics 60. 9/1/86 (1986): 1770-3.

West, Susan. "The new realism: creating an illusion on a computer screen,whether a figure or a cloud, means solving problems of light,form, and texture that every artist faces." Science 84-5 . 7/8/84 (1984): 30-9.

West, Goldberger. "Physiology in fractal dimensions." American Scientist 75. 7,8/87 (1987): 354-65.

Witten, Cates. "Tenous structures from disorderly growth processes." Science 232. 6/27/86 (1986): 1607-12.

Wong, Po-zen. "The statistical physics of sedimentary rock." Physics Today 41. 12/88 (1988): 24-32.

Yadava, R., Bhan, R.K. "Fractal nature of defect clustering in gate oxides of MOS devices." Journal of the Electro-Chemical Society 136. 3/89 (1989): 889-90.

Yang, C., Evesque, P., El-Sayed, M., and . "Effect of variation in the microenvironment of the fractal structure on the donor decay curve resulting from a one-step dipolar energy-transfer process." Journal of Physical Chemistry 90. 3/27/86 (1986): 1284-8.

Yang, C., El-Sayed, M. "Donor-acceptor one-step energy transfer via exchange coupling on a fractal lattice." Journal of Physical chemistry 90. 10/23/86 (1986): 5720-4.

Yang, C., El-Sayed, M., Suib, S. "Apparent fractional dimensionality of uranyl-exchanged zeolites and their photocatalytic activity." Journal of Physical Chemistry 91. 8/13/87 (1987): 440-3.

Yang, C., Chen, Z., El-Sayed, M. "Comparison of the rates of uni- and bimolecular diffusion-controlled reactions on circular filled aggregates and diffusion-limited fractal aggregates in two dimensions." Journal of Physical Chemistry 91. 5/21/87 (1987): 3002-6.

Zorpette, Glenn. "Fractals: not just another pretty picture." IEEE Spectrum 25. 10/88 (1988): 29-31.